R. Boxall

Construction drawing for technicians

Level 1

Longman London and New York

Longman Group Limited
Longman House, Burnt Mill, Harlow
Essex CM20 2JE, England
Associated companies throughout the world

Published in the United States of America
by Longman Inc., New York

First published 1983

British Library Cataloguing in Publication Data
Boxall, R
Construction drawing for technicians level 1.
(Longman technician series)
1. Structural drawing
I. Title
692'.1 TH431

ISBN 0-582-41237-4

Set in 10/11 pt Linotron 202 Times
Printed in Hong Kong by

Longman Technician Series

Construction and Civil Engineering

General Editor – Construction and Civil Engineering

C. R. Bassett, B.Sc., F.C.I.O.B.

Formerly Principal Lecturer in the Department of Building and Surveying, Guildford County College of Technology

Books published in this sector of the series:

Building organisations and procedures *G. Forster*
Construction site studies – production, administration and personnel *G. Forster*
Practical construction science *B. J. Smith*
Construction science Volume 1 *B. J. Smith*
Construction science Volume 2 *B. J. Smith*
Construction mathematics Volume 1 *M. K. Jones*
Construction mathematics Volume 2 *M. K. Jones*
Construction surveying *G. A. Scott*
Materials and structures *R. Whitlow*
Construction technology Volume 1 *R. Chudley*
Construction technology Volume 2 *R. Chudley*
Construction technology Volume 3 *R. Chudley*
Construction technology Volume 4 *R. Chudley*
Maintenance and adaptation of buildings *R. Chudley*
Building services and equipment Volume 1 *F. Hall*
Building services and equipment Volume 2 *F. Hall*
Building services and equipment Volume 3 *F. Hall*
Measurement Level 2 *M. Gardner*
Measurement Level 3 *M. Gardner*
Structural analysis *G. B. Vine*
Site surveying and levelling Level 2 *H. Rawlinson*
Economics for the construction industry *R. C. Shutt*
Design procedures Level 4 *J. M. Zunde*
Design technology Level 5 *J. M. Zunde*
Architectural design procedures *C. M. H. Barritt*
Construction technology for civil engineering technicians *P. L. Monckton*
Environmental science *B. J. Smith, G. M. Phillips and M. Sweeney*
Concrete technology Level 4 *J. G. Gunning*

Contents

Introduction v
Acknowledgements vi
Bibliography iv

Chapter 1	**Materials and equipment**	**1**
Chapter 2	**Drawing procedures**	**41**
Chapter 3	**Plane geometry**	**72**
Chapter 4	**Drawing to scale**	**107**
Chapter 5	**Projections**	**121**
Chapter 6	**Isometric projection**	**148**
Chapter 7	**Oblique projection**	**156**
Chapter 8	**Developments**	**162**
Chapter 9	**The graphical solution of beams**	**179**

Index 200

Introduction

Whichever particular specialization within the construction industry a student ultimately intends to pursue, an understanding of drawing practice and presentation will prove to be not only useful but essential, since virtually all sections of the industry, throughout the entire spectrum of the construction team from site operative to architect, depend significantly on information which is communicated by drawing.

It is true to say that drawing and graphic communication is the principal form of conveying construction information from the point of inception of a project to its completion and beyond. Indeed, a brief insight into the architect's plan of work within the syllabus of Design Procedures will reveal the instrumental part that graphic communication plays in the construction process.

This volume has been prepared primarily for the student of construction, with particular reference to the level 1 TEC Unit Construction Drawing. With this in mind the author has endeavoured to encompass the basic geometrical processes of draughting which are the essential basis of good draughtsmanship, and to extend these processes into a practical type of application.

For those students who wish to pursue a career specifically as draughtsmen or draughtswomen it must be said that any book on draughtsmanship cannot produce instant success. In common with any acquired skill, the route to mastery lies in practice and dedication, which in turn depends largely on a self-motivated desire to improve, experiment and hopefully perfect.

Those students who have convinced themselves that they will never be able to draw should follow the established techniques and methods given in this and other books; it is then certain that the practical tasks and skills will then be simplified. As the tasks become easier and results more acceptable, the skills gained may foster an interest, inspire and motivate the student to discover the true enjoyment that drawing can hold.

It should be appreciated that the ability to draw well reflects benefits throughout a technical course in construction and should not therefore be considered as a subject in isolation. Neatness and presentation are fundamental to all successful studies, in particular subjects such as technology, design procedures and surveying where drawing and sketching form an integral part.

Finally, experience has shown quite conclusively that the standard of drawing achieved by an individual is directly related to his or her degree of interest in the subject. The only able bodied person who cannot draw is one who is disinterested in doing so.

Acknowledgements

We are indebted to the following for permission to reproduce copyright material:

AM Admel, a Division of AM Bruning Ltd for our fig. 1.22; British Standards Institution for extracts from BS.1192, 1969 (complete copies of the standards can be obtained from BSI, 2 Park Street, London WIA 2 BS); Chartpak Ltd for pressure transfers and characters.

The author extends his gratitude to Mrs. Jill Simms for her assistance in preparing the manuscript and to Staedtler (UK) Ltd for trade literature.

Bibliography

Architectural Draughtsmanship, F. Reekie, Edward Arnold.
Architectural Graphics, F. Ching, Architectural Press.
Architects Working Drawing Handbook, Architectural Press.
Building Drawing Practice, BS 1192, British Standards Institution.
Introduction to Structural Mechanics, T. J. Reynolds & L. E. Kent, English Universities Press
Roman Lettering, L. Evetts, Pitman Publishing Ltd.

Chapter 1

Materials and equipment

Drawing boards and associated equipment

Drawing boards may be classified in one of three ways:

1. Boards for use with tee squares.
2. Boards with parallel motion.
3. Board with draughting machine.

All three types should conform to the requirements of BS 4867: Pt 1 : 1972 *Specification for drawing boards and associated equipment*. Boards for use by students should comply with BS 5065 : 1974.

Boards for use with tee square (Figs 1.1, 1.2)

Like all drawing boards this type must maintain a flat true surface for the draughtsman to work upon with a working or datum edge. To ensure minimal movement of the board it is made up of softwood strips, grooved on the back to reduce the tendency to warp, and cross battened with hardwood battens. The battens preserve the true surface but at the same time allow for limited movement in the board by the use of slotted holes at fixing screw positions.

The datum edge usually consists of a piece of hardwood or plastic let into the edge of the board to provide a reliable, straight datum edge, against which the tee square can be operated.

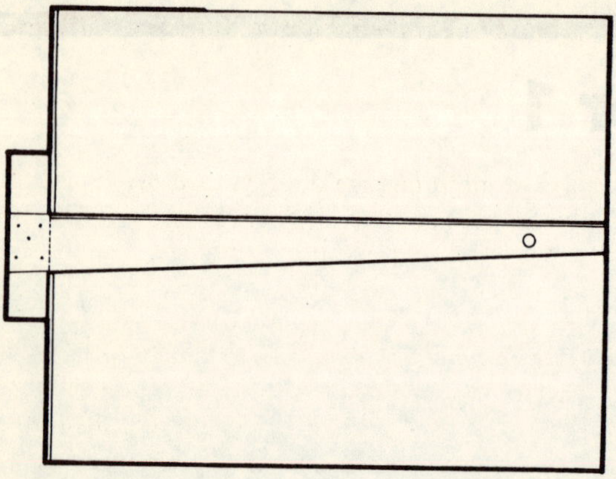

Fig. 1.1 Board with tee square

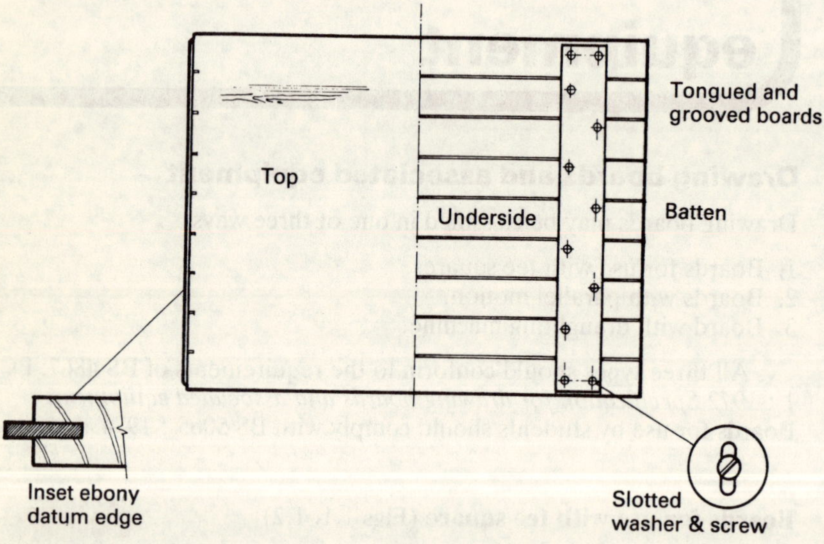

Fig. 1.2

Boards for use with draughting machines and parallel motion units

The type of board used in conjunction with the above equipment can be of the same basic construction to that already described, the only variation being that:

1. A datum edge is not required
   and

2. The board may have a cut-out section at the bottom when used with a parallel motion fitting.
3. The board may incorporate a parking space for equipment when not in use.

In addition to the method of manufacture already described, boards are now made of laminboard, producing a stable and true surface which may be faced with plastic or timber laminate.

Small drawing boards, up to A3 size, are manufactured in high-impact plastic (Fig. 1.3) and incorporate plastic draughting heads and even small draughting machines. They are light and easy to carry and very useful for surveys, etc., but because of their size, and the lack of precision which can be achieved, tend to have a limited application for office use.

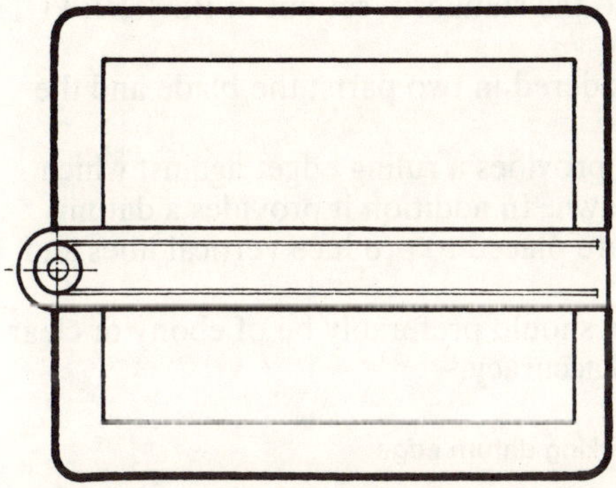

Fig. 1.3 High impact plastic board and square

Care of the board is important. It should be safeguarded from knocks and bumps which may damage the surface or datum edge. It should ideally be situated in a room which is not subject to extremes of heat or moisture which will in turn cause warping. Pins should not be stuck into the timber boards, nor should the board be used as a cutting surface.

A backing sheet should be used on the board surface, as this will provide contrast to line work when drawing on tracing paper; it takes off the hardness of the board surface, as well as eliminating the possibility of the pencil being taken by the direction of the wood grain.

Drawing board sizes

In common with drawing paper, drawing boards are manufactured in a corresponding standard range of sizes. The standard sizes are as shown below.

Board size (mm)		Trimmed paper size (mm)
* A4	277 × 362	297 × 210
* A3	351 × 500	420 × 297
A2	650 × 470	594 × 420
A1	920 × 650	841 × 594
A0	1270 × 920	1189 × 841

* Special boards with built-in squares

Straight edges

Tee squares (Fig. 1.4)

Tee squares are manufactured to the standards set out in BS 4867: Pt 2: 1972 and BS 5065: 1974.

The tee square may be considered in two parts; the blade and the stock or butt.

The blade of the tee square provides a ruling edge, against which all horizontal lines should be drawn. In addition it provides a datum edge against which set squares are placed to produce vertical lines and the incremental angles of 15°.

The ruling edge of the blade should preferably be of ebony or clear plastic to provide durability and accuracy.

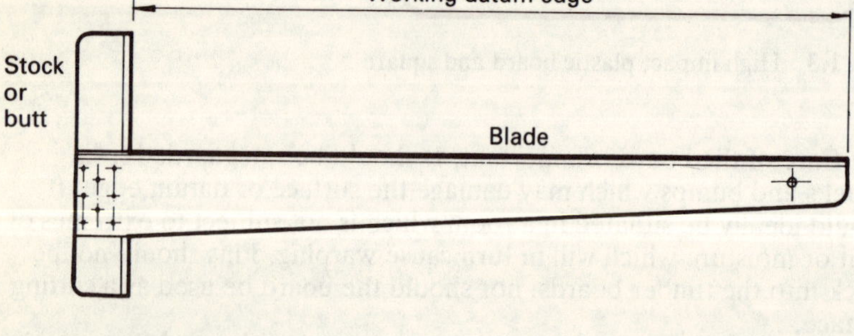

Tee square sizes

Size	Working edge length
A_2	650 mm
A_1	920 mm
A_0	1270 mm

Fig. 1.4

The stock of the tee square is screwed rigidly to the blade, making an angle of 90° with the ruling edge. The stock is pressed firmly against the datum edge of the board when in use to ensure that consistent horizontal and vertical lines are maintained.

Careless handling of a tee square will render it useless. The ruling edge should never be used as a cutting edge and the temptation to use the square as a hammer must be resisted. Cleanliness is important; as the blade of the square is constantly moving over the drawing, dust and dirt will be spread over the surface from the underside of the blade.

Left-handed squares are available, the stock operating against the right-hand side of the board as opposed to the left.

Parallel motions (Fig. 1.5)

The parallel motion mechanism provides the draughtsman with a smooth operating horizontal blade which moves vertically over the drawing on steel cables.

The blade can be adjusted on the cables, but once secured in position the drawing edge cannot move out of square.

The blade is made from clear plastic or hardwood and often incorporates a rest for pens and pencils. It is secured at its ends to cables which pass over helically grooved wheels at the top edge of the board, and single grooved wheels at the bottom edge. To maintain the smooth action and balance of the blade, counter balance weights are fixed to the cables. This type of board arrangement is usually used in conjunction with a drawing board stand.

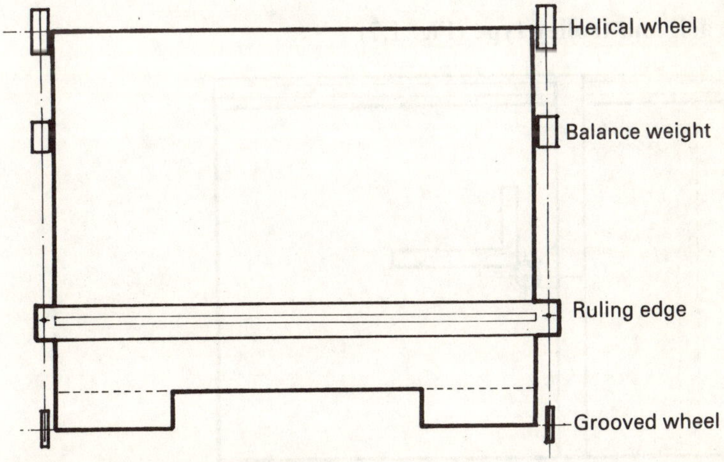

Fig. 1.5 Board with parallel motion

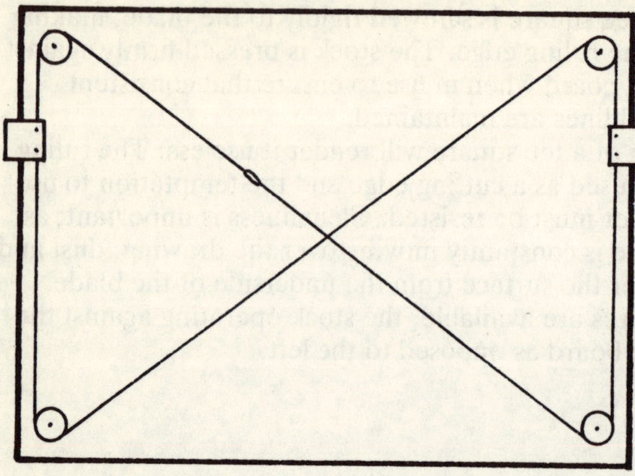

Fig. 1.6 Cross wired parallel motion

An alternative to this type of parallel motion is shown in Fig. 1.6. Here the wires are run around four corner pullies, and cross at the centre of the board. It has the advantage that the mechanism is contained in the size of the board, and no counter balance weights are required.

Draughting heads can be fitted to parallel motion blades to provide a simple form of draughting machine.

Draught machines

Draughting machines fall into two categories.

(a) Track and trolley type (Fig. 1.7)

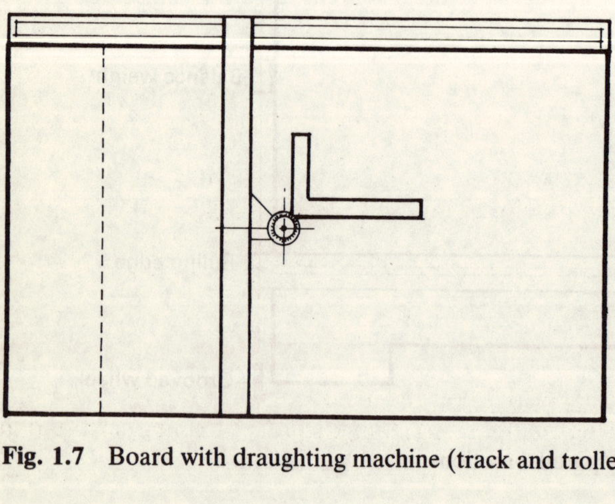

Fig. 1.7 Board with draughting machine (track and trolley type)

This has a vertical bar, precision engineered to run on a horizontal sliding track across the drawing board. The vertical bar provides a constant and reliable vertical datum, fixed to which is a draughting head which in turn slides vertically up and down the bar.

Draughting techniques may have to be adapted to suit this type of machine, as long lines are not so easily drawn with the relatively short blade length of the draughting head. Despite this, draughting machines of this type are pleasant and efficient in use.

(b) Parallelogram type (Fig. 1.8)

This type consists of a mobile arm moving in parallel fashion across the board in all directions about a fixed point. A drafting head is fixed to the end of the arm.

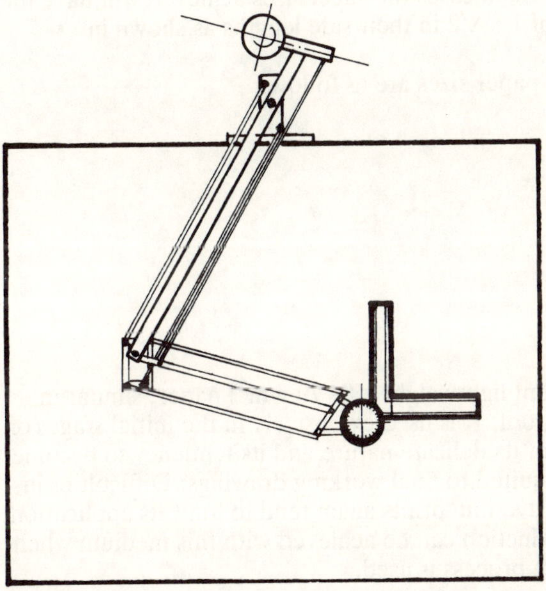

Fig. 1.8 Board with draughting machine (parallelogram type)

Draughting machines are now available with draughting heads incorporating L.C.D. digital read-out display for angular measurement.

Papers and films

Paper sizes

Standardization in paper sizes results in economies not only in the drawing paper, but also in printing, filing, storage, mailing, etc. and in the convenience of matching paper size to board size.

The accepted method of standardization adopted in this country is the international A series of sizes.

The basis of the A sizes is the A0 sheet (Fig. 1.9a) which has a surface area of 1 m^2 with its sides in the ratio of $1 : \sqrt{2}$, for A0 size, the sheet side lengths being 841 × 1189 mm,

$$\text{i.e.} \quad y = x\sqrt{2} \qquad \therefore \quad y = 841\sqrt{2} = 1189 \text{ mm}$$

$$\text{and} \quad x = \frac{y}{\sqrt{2}} \qquad \therefore \quad x = \frac{1189}{\sqrt{2}} = 841 \text{ mm}$$

and the sheet area =

$$1189 \times 841 \text{ mm} = 0.999\,949 \text{ m}^2 \ (1 \text{ m}^2)$$

Further sizes in the A series are obtained by halving the sheet across its longest side. In all cases the sheet sizes achieved will have the constant relationship of $1 : \sqrt{2}$ in their side lengths as shown in Fig. 1.9b.

Standard trimmed paper sizes are as follows:

A0 1189 × 841 mm
A1 841 × 594 mm
A2 594 × 420 mm
A3 420 × 297 mm
A4 297 × 210 mm

Types of paper

Detail paper
This is a semi translucent lightweight (50–70 g/m^2) paper, similar in appearance to greaseproof. It is used extensively in the initial stages of a design, but because of its delicate nature and its tendency to become brittle with age, is not suited to final working drawings. Difficulties in producing satisfactory dye-line prints again tend to limit its application, although quality reproduction can be achieved with this medium when an electrostatic copying process is used.

Tracing paper
This is probably the most extensively used medium in practice, because of its economy, its suitability to dye-line reproduction and its smooth surface which is suitable for either ink or pencil drawing.

Two types of tracing paper are available: *natural*, the most extensively used, and *prepared*, which is less used because of its expense, but is more durable although less translucent than natural.

Tracing papers are dimensionally unstable and variation in moisture content can cause significant dimensional change. For this reason dimensions should not be scaled from such drawings. Care must be exercised when using tracing papers as they can be easily damaged, particularly when erasing ink lines with a razor blade.

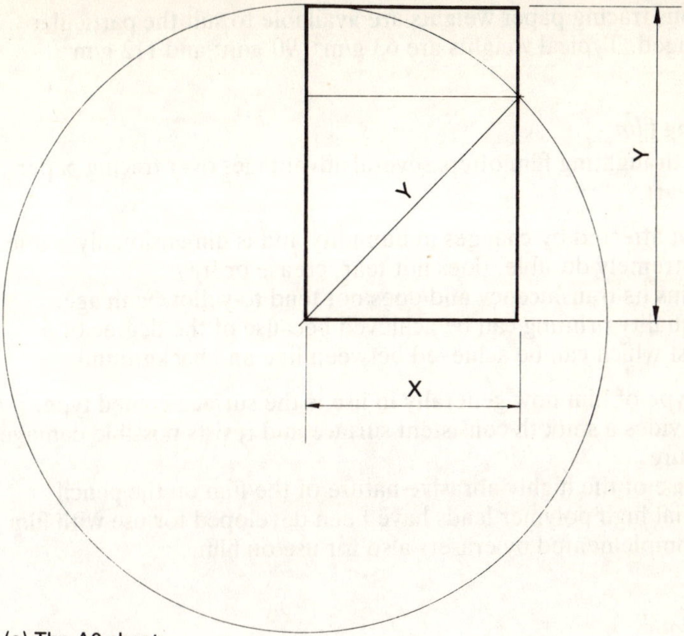

(a) The A0 sheet

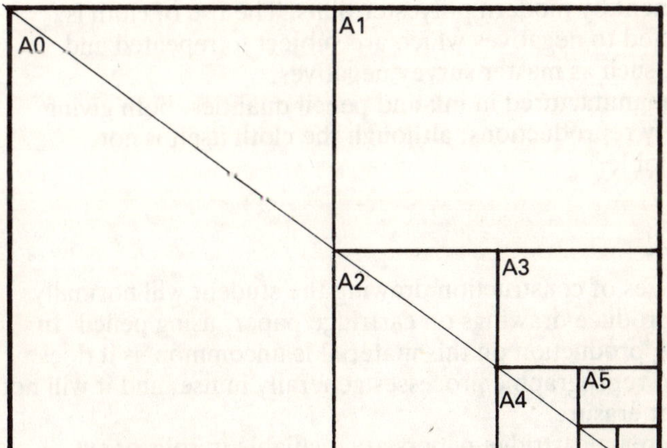

(b) The A series of paper sizes

Fig. 1.9

Various tracing paper weights are available to suit the particular use envisaged. Typical weights are 63 g/m^2, 90 g/m^2 and 112 g/m^2.

Draughting film

Polyester draughting film offers several advantages over tracing paper, in particular:

1. It is not affected by changes in humidity and is dimensionally stable.
2. It is extremely durable, does not tear, crease or fray.
3. It retains its translucency and does not tend to yellow with age.
4. High quality printing can be achieved because of the degree of contrast which can be achieved between line and background.

The type of film now generally in use is the surface-coated type, which provides a smooth consistent surface and resists possible damage from erasure.

Because of the highly abrasive nature of the film on the pencil lead, special high polymer leads have been developed for use with film and are complemented by erasers also for use on film.

Tracing cloth

This is a specially treated high-quality linen material, having semi-translucent properties, making it suitable for dye-line reproduction. In practice cloth is now little used because of its expense and its replacement by modern polyester films. The use of cloth is normally restricted to negatives which are subject to repeated and continuous use, such as master survey negatives.

Cloths are manufactured in ink and pencil qualities, both giving very good quality reproductions, although the cloth itself is not dimensionally stable.

Cartridge paper

In the initial stages of construction drawing the student will normally be required to produce drawings on cartridge paper, using pencil. In practice drawing production on this material is uncommon as it does not lend itself to reprographic processes generally in use, and it will not tolerate constant erasure.

White and cream cartridge papers are available in rolls or cut sheets, and are manufactured in three weights – 'thin', 'medium' or 'stout'. Student grade paper is ideally suited for work in pencil, ink work being less satisfactory. It should be appreciated that cartridge paper has a 'right' and 'wrong' side, the wrong side having a slightly rough and pitted surface which can be identified by careful inspection. The cut edge of trimmed sheets usually turns down toward the wrong side.

Set squares

Set squares are manufactured in clear or tinted plastic, and are used for drawing vertical and inclined lines against the datum edge of the tee square. There are three common types of set squares (Fig. 1.10): the 60–30°, the 45° and the adjustable. Used together the 60–30° and 45° squares can be used to draw angles from 15° to 90° in increments of

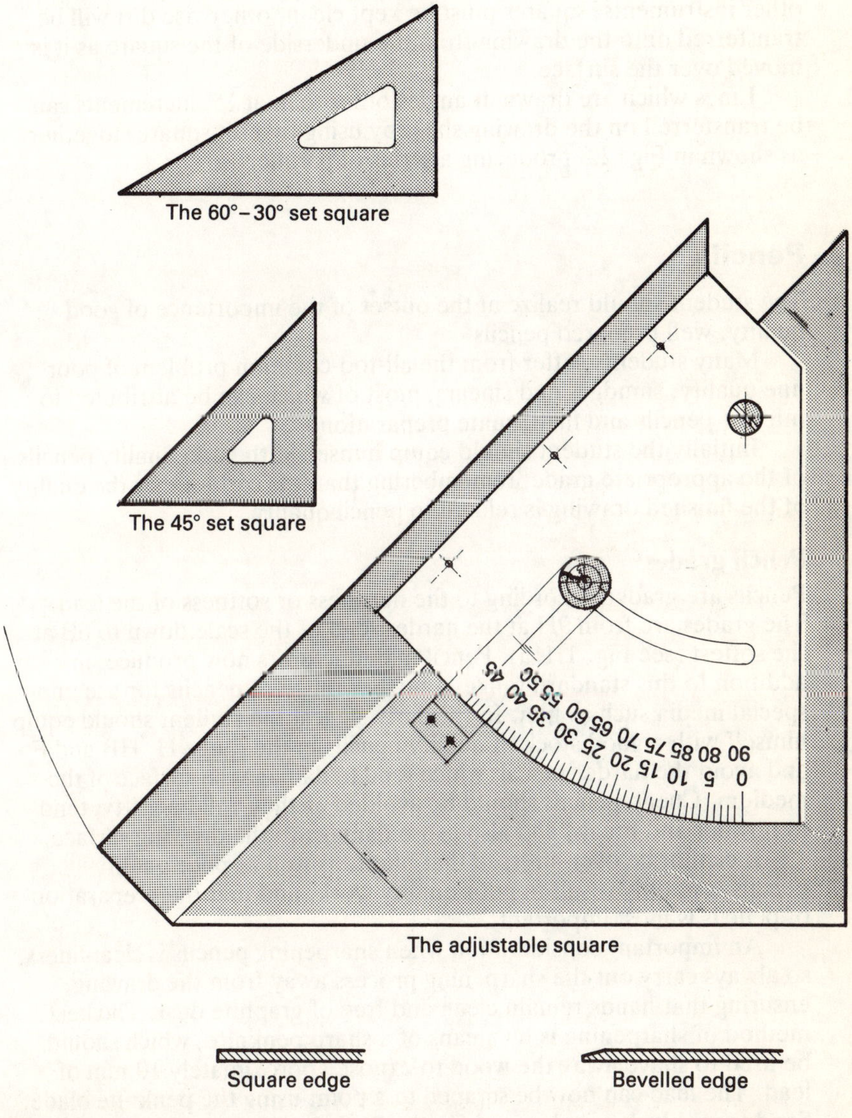

Fig. 1.10 Set squares

15° (see Ch. 3). The adjustable has the advantage that it can be set to any required angle obviating the need for a protractor.

Initially the student should acquire squares having a longest side of about 225 mm and preferably with a square edge as opposed to a bevelled edge. The square edge reduces the tendency for ink to smudge when ink lines are drawn.

Care of set squares is important. The edges must be carefully preserved and should not be used as a cutting straight edge. As with other instruments, squares must be kept clean, otherwise dirt will be transferred onto the drawing from the underside of the square as it is moved over the surface.

Lines which are drawn at angles other than at 15° increments can be transferred on the drawing sheet by using two set squares together as shown in Fig. 2.3 producing a series of parallels.

Pencils

The student should realize at the outset of the importance of good quality, well prepared pencils.

Many students suffer from the all-too-common problem of poor line quality, smudges and smears, most of which can be attributed to inferior pencils and inadequate preparation.

Initially the student should equip himself with good quality pencils of the appropriate grade, remembering that it is certain that the quality of the finished drawing is related to pencil quality.

Pencil grades

Pencils are graded according to the hardness or softness of the lead. The grades are from 9H at the hardest end of the scale down to 6B at the softest (see Fig. 1.11a). Pencil manufacturers now produce, in addition to this standard range, a further range of pencils for use on special media such as film. In the early stages, the student should equip himself with a good stock of quality pencils in the range H, HB and F and avoid the harder pencils which tend to damage the surface of the medium. Cheap pencils should be avoided as they will be gritty, tend to powder and break, and also cause damage to the drawing surface.

It cannot be over-stressed that blunt, as well as poor quality pencils, can only result in poor quality work, and that the preparation of pencils is most important.

An important consideration when sharpening pencils is cleanliness, so always carry out the sharpening process away from the drawing, ensuring that hands remain clean and free of graphite dust. The best method of sharpening is by means of a sharp penknife, which should be used to shave away the wood to expose approximately 10 mm of lead. The lead can now be scraped to a point using the penknife blade. Sandpaper blocks can be used for the final sharpening of the lead, but

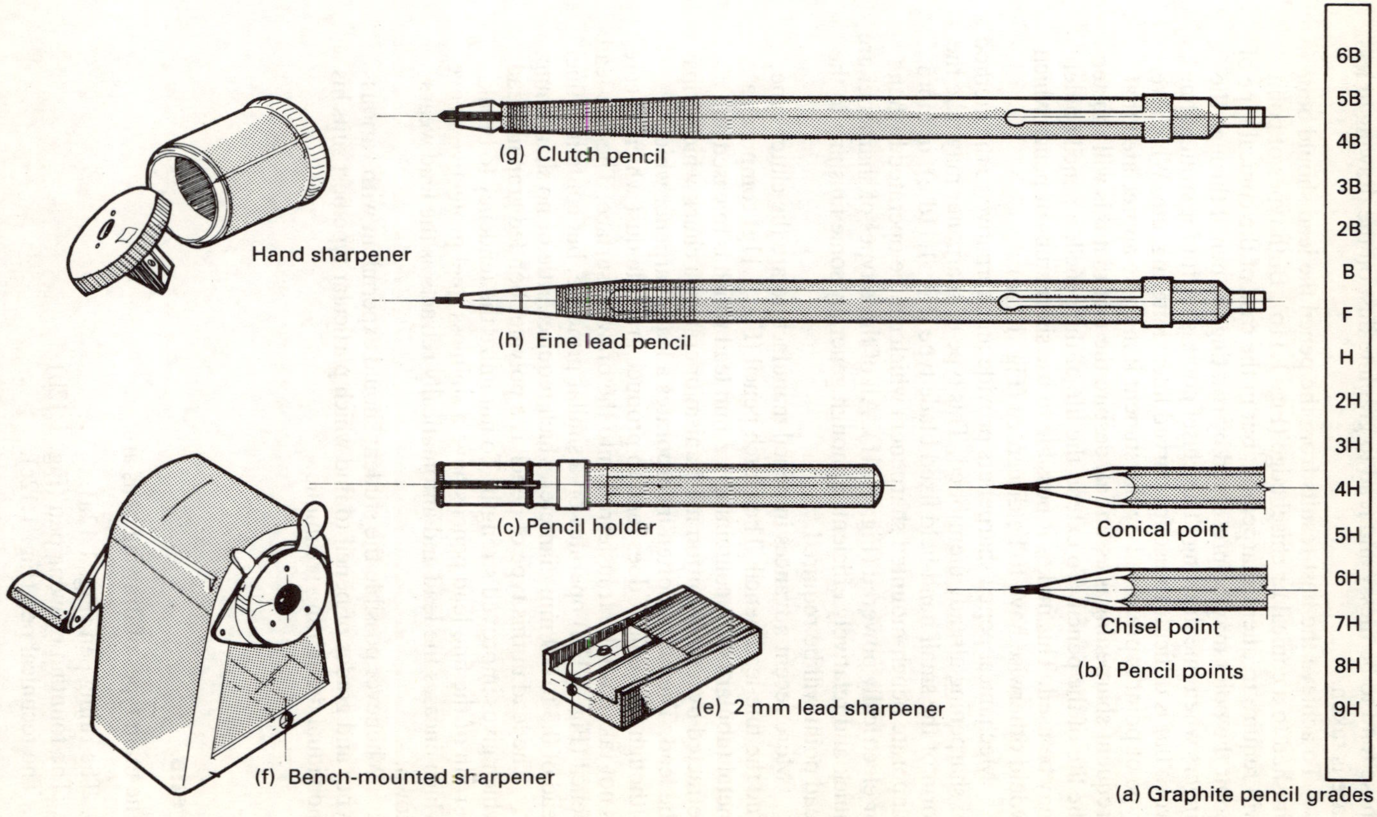
6B
5B
4B
3B
2B
B
F
H
2H
3H
4H
5H
6H
7H
8H
9H
(a) Graphite pencil grades
Conical point
Chisel point
(b) Pencil points
(c) Pencil holder
(e) 2 mm lead sharpener
(f) Bench-mounted sharpener
(g) Clutch pencil
(h) Fine lead pencil
Hand sharpener

Fig. 1.11

these tend to be messy and lead to graphite dust on the drawing unless care is taken.

To achieve the best results from the pencil the lead should be finished to a conical or chisel edge (Fig. 1.11b). Both types of point will require frequent sharpening, but in the case of the conical type of point the point may be preserved by rotating the pencil through the fingers when line drawing. The chisel point is useful for straight line work but is difficult to use for curved lines, lettering, etc. Whichever type of point is preferred by the student it must be appreciated that frequent sharpening is essential, despite the fact that this will shorten the life of the pencil. To extend the life of the pencil, a pencil holder may be used. This makes it possible to use short stubs of pencil which would otherwise have to be discarded (Fig. 1.11c).

Mechanical pencil sharpeners provide an alternative to the method of sharpening already mentioned. This type of sharpener may take the form of the small hand-held fixed blade type (Fig. 1.11d,e), or more elaborate bench-mounted sharpeners which can be operated by hand or electrically powered (Fig. 1.11f). All of these types of sharpener are quick and relatively efficient, although generally some finishing to the lead point will be required.

More recent advances in pencil manufacture are the clutch type and the fine-lead pencil. The clutch pencil (Fig. 1.11g) comprises a metal tubular body, containing a 2 mm lead which is released and retracted by a push-button mechanism operating a chuck which grips the lead. The button often incorporates a lead sharpener which although convenient does tend to produce graphite dust which, if care is not taken will be transferred onto the drawing surface. The fine-lead pencil (Fig. 1.11h) operates on a similar principle, but contains a finer lead of 0.3 to 0.9 mm diameter which requires little or no sharpening.

The lead in this type of pencil is a polymer bonded graphite lead which gives improved strength. To minimize the tendency to break, certain of the fine lead pencils have a stainless steel protective sleeve which encases the lead and automatically retracts as the lead wears down.

Whenever possible the student should experiment with various types and grades of pencil to find which particular type best suits his individual draughting technique.

Pens

The main types of drawing pens are:

1. The ruling pen (Fig. 1.12a).
2. The fountain drawing pen (Fig. 1.12b).
3. The technical pen (Fig. 1.12c).

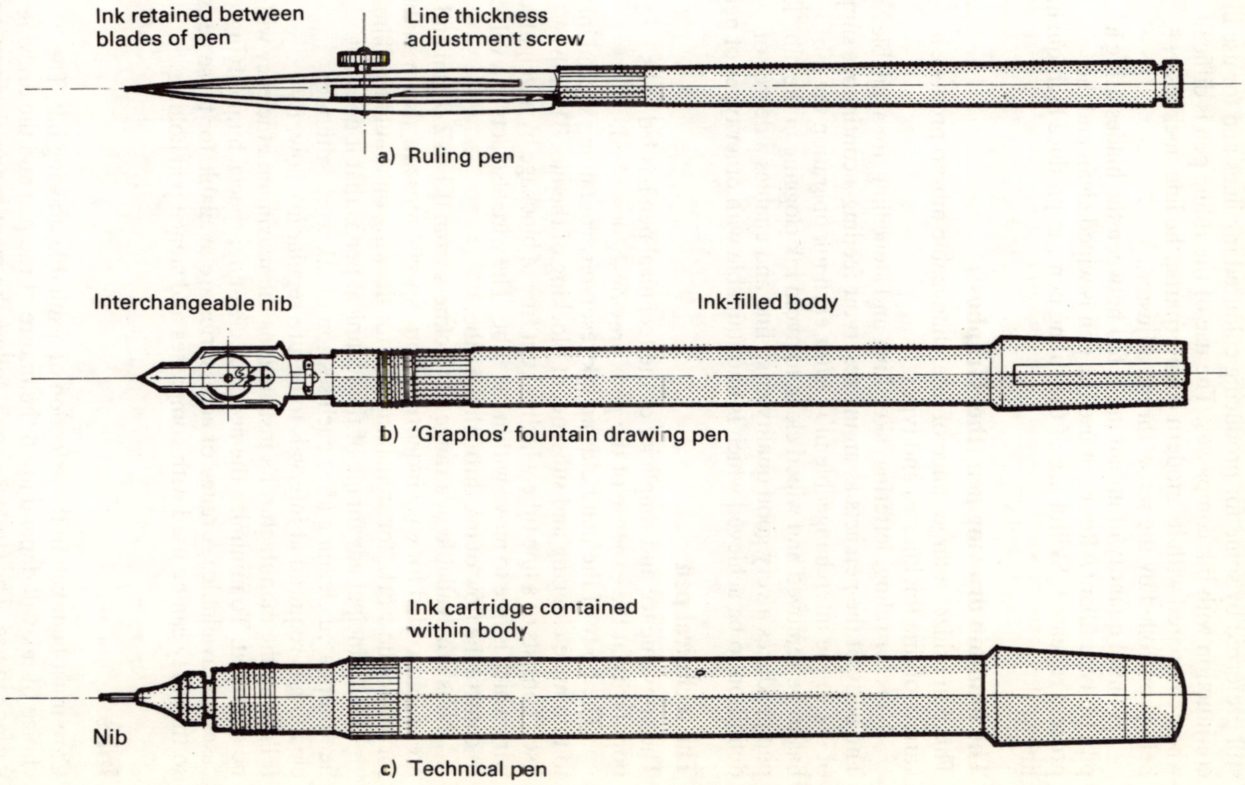
Ink retained between blades of pen
Line thickness adjustment screw
a) Ruling pen
Interchangeable nib
Ink-filled body
b) 'Graphos' fountain drawing pen
Ink cartridge contained within body
Nib
c) Technical pen

Fig. 1.12

The ruling pen

Although now largely replaced by the technical pen, the ruling pen is still preferred by some for producing coloured ink lines and for use in conjunction with ink compasses. The use of the ruling pen is difficult when compared with its modern counterparts, the line thicknesses being controlled by the screw thread adjuster.

A small quantity of indian ink is fed between the blades using a pipette type filler. After each use the pen is wiped clean and the process repeated. Skilful use of the ruling pen can produce high-quality linework.

The fountain drawing pen (the 'Graphos')

This pen utilizes a large range of interchangeable nibs to produce a variety of line thicknesses and types.

Nibs for ruling, lettering, sketching and stencilling are available. The body of the pen acts as an ink reservoir feeding a continuous supply of ink to the interchangeable nib. After each ink operation the nib has to be removed and wiped clean to avoid ink clogging in the nib. This pen produces a very good quality crisp line and enables a degree of definition to be achieved which is unobtainable with other types of pen.

The technical pen

The convenience and simplicity of the technical pen has led to its popularity at the expense of the pens previously described.

The body of the pen holds an ink reservoir, containing special ink to minimize clogging and subsequent blocking of the nib. The nib itself consists of a fine tube which is kept free of blockage by a weighted wire which is free to move within the tube. The wire also acts as a valve, to control the flow of ink through the tube.

Nibs are available in a range of thickness from 0.1 to 2.0 mm and are colour coded for easy indentification. Special nibs are also available for microfilm work, for stencilling and for drawing on draughting film.

The principal advantage of the technical pen is that it does not need repeated cleaning after each session of ink work, although cleaning is required at intervals to ensure regular ink flow. By following the manufacturer's instructions the maximum efficiency will be obtained. To minimize the need for cleaning, special humidifying cases are available. A range of accessories are available for these pens so that they can be used with compasses and stencil pen holders.

Inks

Care must be taken in the selection of a suitable drawing ink. The choice of ink will depend upon the nature of the pen and the drawing surface in use. The student should follow the manufacturer's directions when selecting inks for technical pens or fountain pens to avoid damage to the instrument.

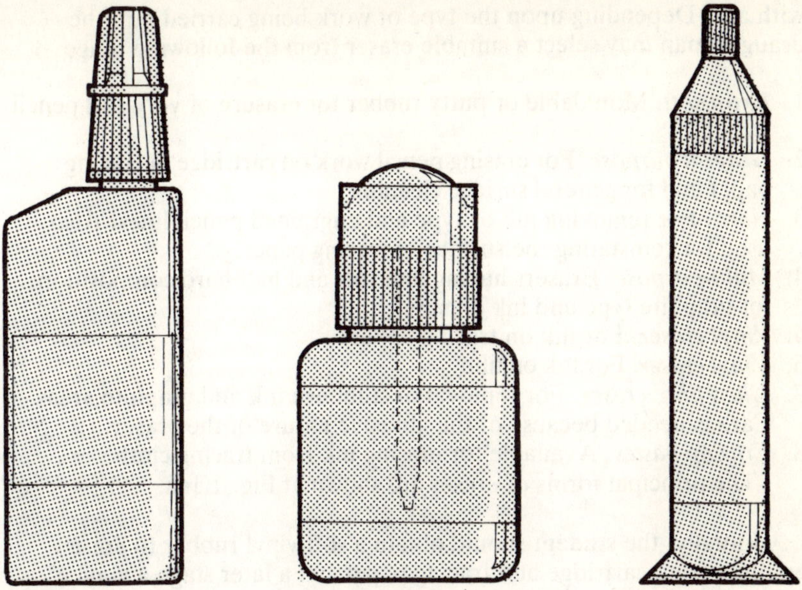

Fig. 1.13 Ink dispensers

Inks used on tracing films have an etching quality which may prove harmful unless the correct pen is employed.

Ink quality does vary from manufacturer to manufacturer and a certain amount of experimentation might prove worthwhile at a later stage. For example waterproof drawing inks may be diluted with distilled water to produce a less dense line if required, or to aid the flow of the ink in hot, dry conditions.

Manufacturers now market inks in glass bottles with pipette or dropper lids, in cartridges, or in plastic squeeze bottles (Fig. 1.13) and are available in a range of colours. The use of coloured inks on tracing paper is not generally recommended, however, because of the poor definition achieved in subsequent reprographic processes.

Erasers

Just as quality in pencils is important to the draughtsman, so is the quality, selection and use of erasers. Modern draughting erasers tend to be a plastic– or vinyl-based material, produced in various types for specific uses. These erasers are superior in several respects to the conventional type of rubber in that they are generally cleaner and more efficient in use, and do not have the tendency to surface-harden with age. Depending upon the type of work being carried out, the draughtsman may select a suitable eraser from the following range.

1. *Extra soft*. Mouldable or putty rubber for erasure of very soft pencil work.
2. *General purpose*. For erasing pencil work on cartridge or tracing paper and for general surface cleaning.
3. *Hard*. For removing ink or type and engrained pencil lines. Also used for reinstating the surface of tracing paper.
4. *Multi-purpose*. Erasers having half soft and half hard composition for graphite type and ink erasure.
5. *Vinyl eraser*. For ink on tracing paper.
6. *Vinyl eraser*. For ink on film.
7. *Glass fibre eraser*. For removal of stubborn ink and graphite areas. Care is needed because of the abrasive nature of the eraser.
8. *Erasing pastes*. Available for erasing ink from tracing cloth.

The principal forms of eraser are shown at Fig. 1.14a.

Initially, the student should obtain a soft vinyl rubber to erase graphite from cartridge and tracing paper. At a later stage a vinyl rubber may be of use to erase ink from tracing paper. Erasers of this type contain a chemical solvent which acts upon the ink work without damaging the drawing surface.

An alternative method of ink erasure from tracing paper is by scratching away the ink line with a razor blade and then reinstating the surface of the drawing with a hard eraser. The reinstatement of the surface prevents ink absorbtion and spreading when the line is redrawn.

Electric erasers (Fig. 1.14b)

In large drawing offices, electrically powered erasers are often installed. The eraser consists of a hand-held power tool, fitted with an eraser plug which is rotated at high speed by the electric motor. This type of eraser can be accurately directed at the area to be erased and is useful where large areas of dense ink have to be removed. Care must be exercised to avoid serious damage to the drawing.

The erasing shield (Fig. 1.14c)

It is often necessary to erase only a particular portion of a drawing and in doing so the draughtsman may wish to preserve adjacent line work. To do this the erasing shield may be used. The shield is generally made from thin stainless steel sheet, with assorted openings through which errors can be erased when the shield is placed on the drawing surface. Careful use of the shield will protect adjoining linework and minimize the need for reinforcing existing line work.

In a continual effort to keep the drawing clean, particles left on the drawing after the erasing operation should be carefully brushed clear away from the surface with a brush or soft cloth.

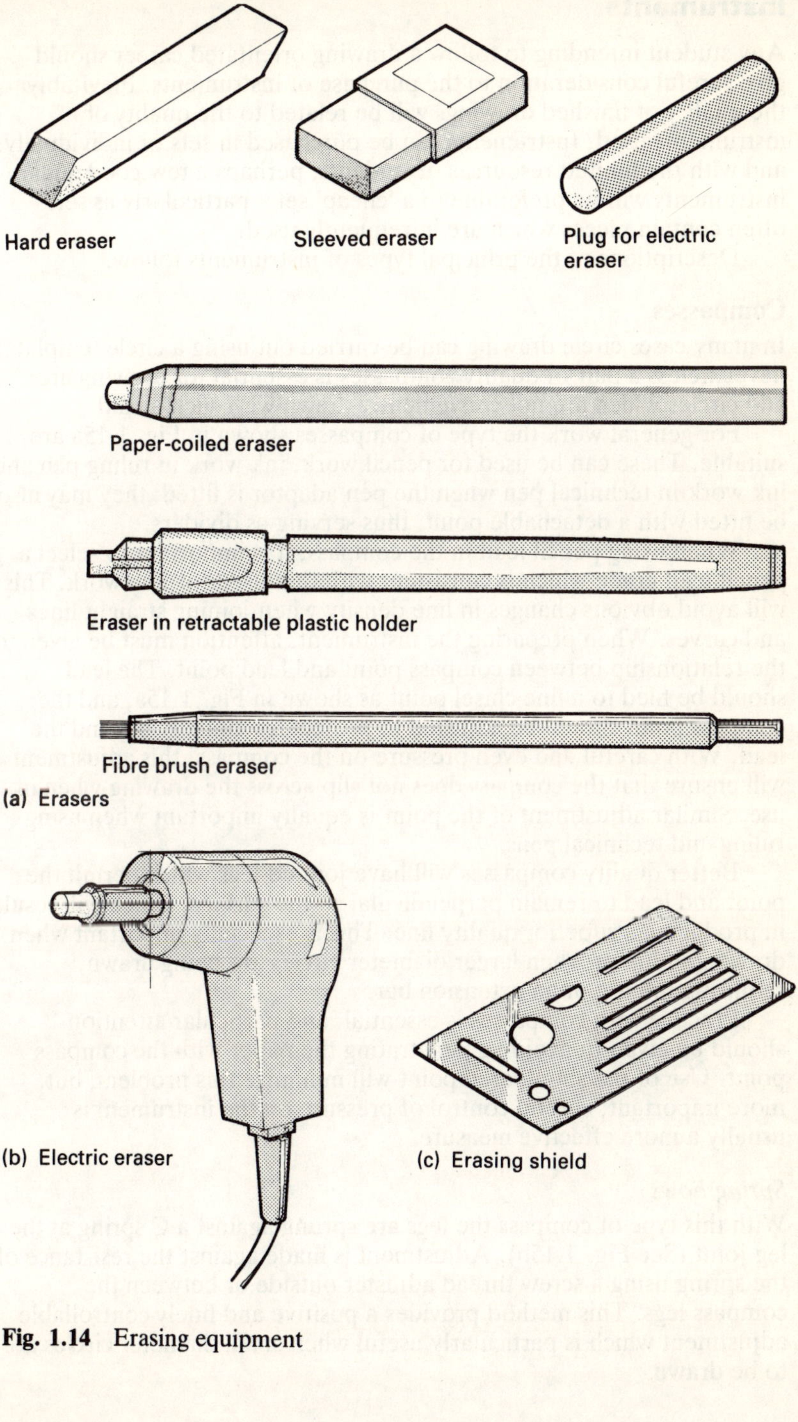

Fig. 1.14 Erasing equipment

Instruments

Any student intending to follow a drawing orientated career should give careful consideration to the purchase of instruments. Inevitably the quality of finished drawings will be related to the quality of instruments used. Instruments can be purchased in sets or individually, and with the limited resources of students, perhaps a few good quality instruments will be preferable to a 'cheap' set – particularly as sets often contain pieces which are infrequently used.

Descriptions of the principal types of instruments follow.

Compasses

In many cases circle drawing can be carried out using a circle template; nevertheless a pair of quality compasses is essential for drawing arcs and circles which are not conveniently drawn with such an aid.

For general work the type of compasses shown in Fig. 1.15a are suitable. These can be used for pencil work, ink work in ruling pen and ink work in technical pen when the pen adaptor is fitted; they may also be fitted with a detachable point, thus serving as dividers.

When using pencil lead in the compass, it is important to select a pencil lead grade which is consistent with that used for line work. This will avoid obvious changes in line density when joining straight lines and curves. When preparing the instrument, attention must be given to the relationship between compass point and lead point. The lead should be filed to a fine chisel point as shown in Fig. 1.15a, and the compass point should be adjusted to project fractionally beyond the lead. With careful and even pressure on the compass, this adjustment will ensure that the compass does not slip across the drawing when in use. Similar adjustment of the point is equally important when using ruling and technical pens.

Better quality compasses will have jointed legs which permit the point and lead to remain perpendicular to the surface, which will result in producing a superior quality line. This is especially important when drawing in ink or when larger diameter circles are being drawn, requiring the use of an extension bar.

Practice with compasses is essential, and particular attention should be given to avoiding penetrating the paper with the compass point. Use of the shouldered point will minimize this problem, but, more important, careful control of pressure on the instrument is usually a more effective measure.

Spring bows

With this type of compass the legs are sprung against a C spring at the leg joint (See Fig. 1.15b). Adjustment is made against the resistance of the spring using a screw thread adjuster outside or between the compass legs. This method provides a positive and finely controllable adjustment which is particularly useful when small diameter circles are to be drawn.

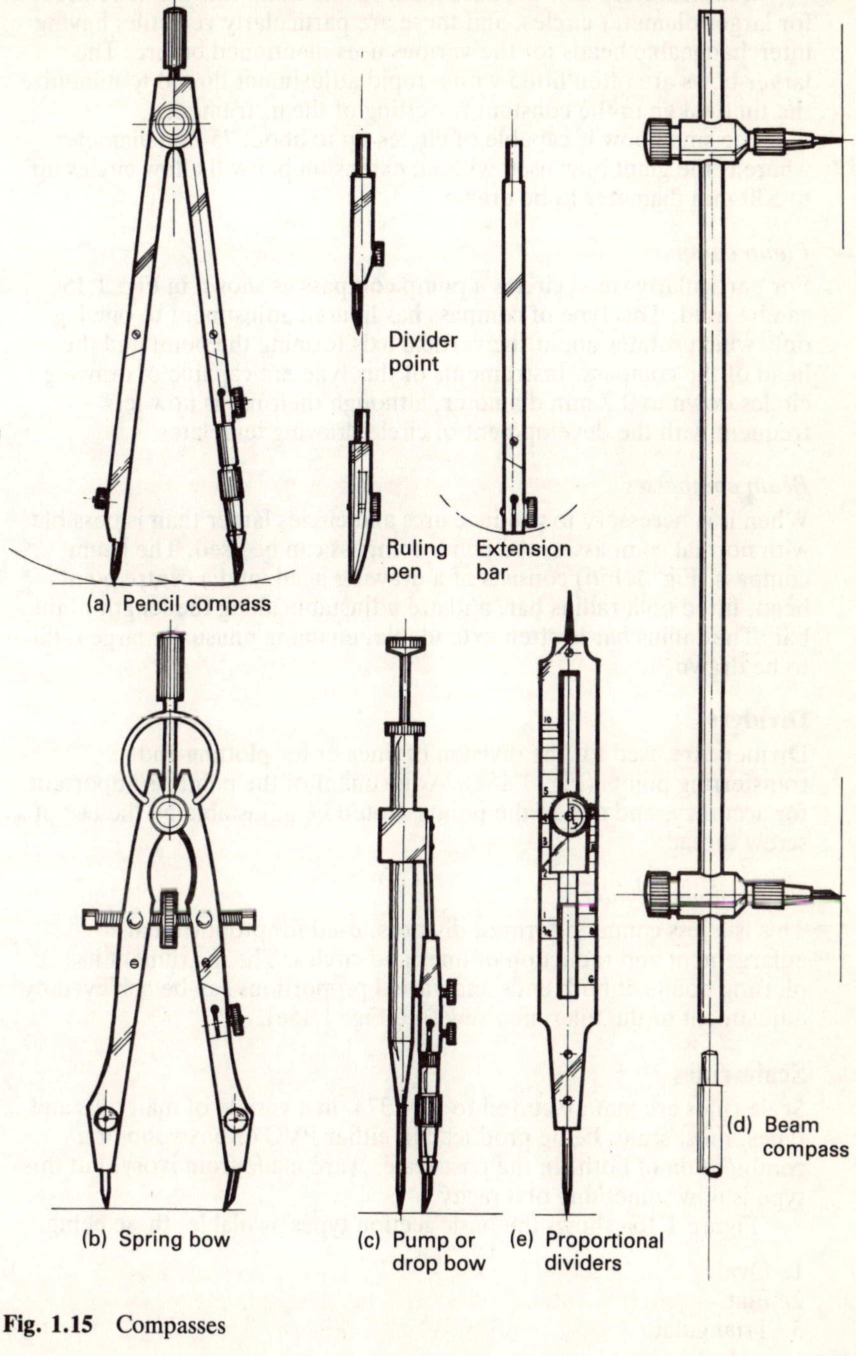

Fig. 1.15 Compasses

Manufacturers now produce giant spring bows which can be used for larger diameter circles, and these are particularly versatile, having interchangeable heads for the various uses mentioned before. The larger bows are often fitted with a rapid adjustment thread to minimize the time taken in the constant re-setting of the instrument.

The small bow is capable of circles up to about 75 mm diameter, whereas the giant bow used with an extension bar will allow circles up to 530 mm diameter to be drawn.

Pump compass

For particularly small circles a pump compass as shown in Fig. 1.15c can be used. This type of compass has limited adjustment to one leg only which rotates about the vertical axis forming the point and the head of the compass. Instruments of this type are capable of drawing circles down to 0.7 mm diameter, although their use is now less frequent with the development of circle drawing templates.

Beam compasses

When it is necessary to produce arcs and circles larger than is possible with normal compasses, the beam compass can be used. The beam compass (Fig. 1.15d) consists of a drawing head, and a centre point head, fitted on a radius bar, and are adjustable along the length of the bar. The radius bar is often extendable, enabling unusually large radii to be drawn.

Dividers

Dividers are used for the division of lines or for plotting and transferring points (Fig. 1.15c). Adjustment of the points is important for accuracy, and one of the points should be adjustable by the use of a screw thread.

Proportional dividers

This is a less common form of dividers, used for proportional enlargement and reduction of lines and circles. The instrument has plotting points at both ends and varied proportions can be achieved by adjustment of the fulcrum screw (see Fig. 1.15e).

Scale rules

Scale rules are manufactured to BS 1374, in a variety of materials and types, most scales being produced in either PVC or boxwood or a combination of both. In the past scales were made from ivory, but this type is now something of a rarity.

Figure 1.16a shows the basic section types available, these being:

1. Oval.
2. Flat.
3. Triangular.

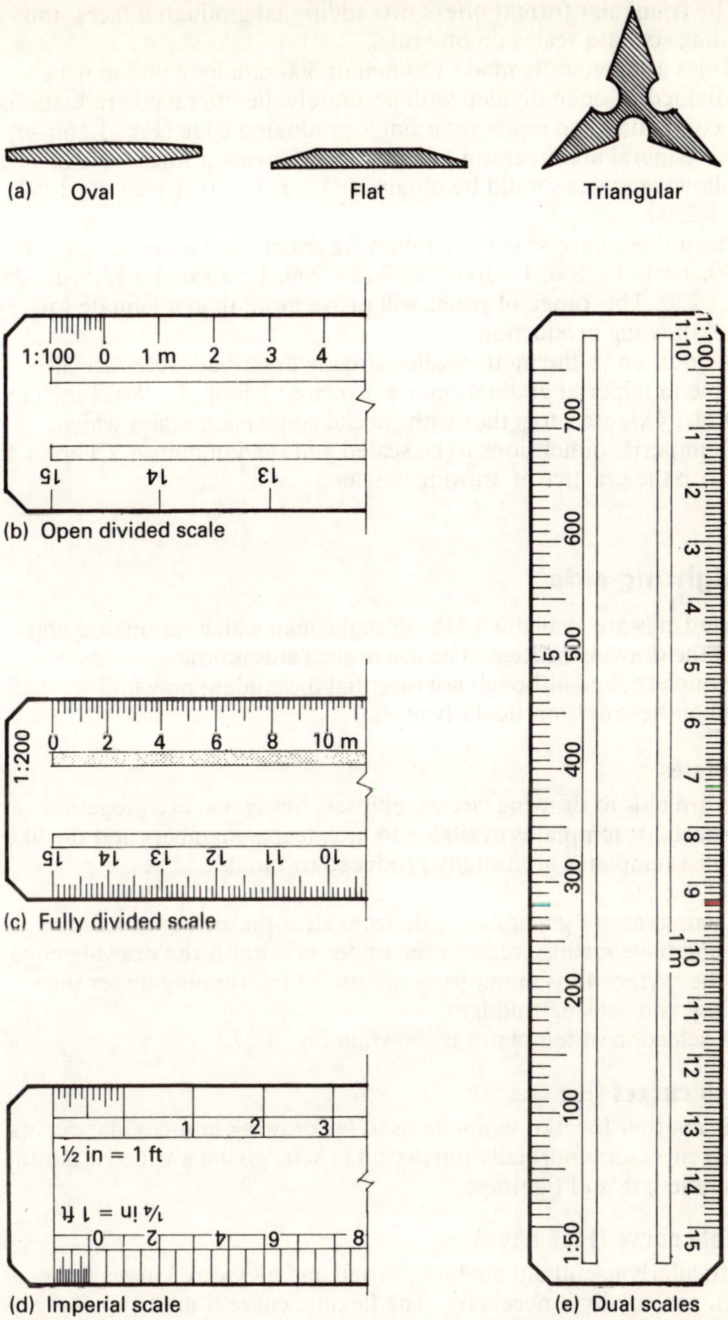

Fig. 1.16 Scale rules

The triangular format offers two additional graduated faces, thus providing six base scales on one rule.

Rules are generally made 150 mm or 300 mm long and may be fully divided or open divided with accurately die-engraved graduations, often combining two scales on a single graduated edge (Fig. 1.16b–e).

For general architectural and building drawing a single rule having the following scales should be obtained: 1 : 5, 1 : 10, 1 : 20, 1 : 1250 and 1 : 2500.

From these base scales the following scales can be derived: 1 : 50, 1 : 500, 1 : 1, 1 : 100, 1 : 1000, 1 : 2, 1 : 200, 1 : 2000, 1 : 12.5, 1 : 25 and 1 : 250. This range of scales will prove more than adequate for general drawing production.

In addition to the metric scales already described, rules are also available in imperial graduation, i.e. $\frac{1}{8}$ inch to 1 foot (1 : 96), $\frac{1}{4}$ inch to 1 foot (1 : 48), etc., together with special conversion scales which enable imperial dimensions to be scaled and read in metric. Chapter 4 deals with the practice of drawing to scale.

Draughting aids

Assorted aids are available to the draughtsman which rationalize and hasten the drawing process. The use of such aids is now commonplace, and although not essential the student may find certain of these aids particularly useful.

Templates

These are aids to drawing circles, ellipses, hexagons, etc., together with specialist templates available to architects, engineers and the like. Specialist templates are usually produced to suitable scales, e.g. 1 : 100, 1 : 50.

Templates are generally made from clear plastic about 1.5 mm thick and have raising studs on the under side to lift the drawing edge from the surface thus minimizing the risk of ink running under the template and causing smudges.

A selection of templates is shown in Fig. 1.17.

French curves (Fig. 1.18a)

This is another form of template used for drawing non-circular curves. French curves are normally purchased in sets, giving a wide variation in curve length and tightness.

Flexible curve (Fig. 1.18b)

A particularly useful aid enabling varied and awkward curves to be drawn, repeatedly if necessary. The flexible curve is made up of a plastic-coated steel core which will tolerate bending, and once bent will hold its form until readjusted.

Railway curves (Fig. 1.18c)
These have a specialist application in that they are particularly used for drawing the large radius curves encountered in railway layout and design or in naval architecture.

Splines (Fig. 1.18d)
This is a thin strip of spruce or ply which can be laid onto the drawing and held to the required curve using lead weights.

Pressure transfers

Lettering, symbols, renderings, textures and title blocks can be applied to drawings in the form of a direct transfer.

The transfer system consists of a transparent plastic carrier film onto which vinyl ink characters are printed. The characters are protected from damage and accidental transfer by a silicone-treated backing sheet.

To apply a transfer, the backing sheet must be removed and the film accurately positioned on the drawing surface. Provided the surface is smooth, clear and free from grease, the character can be applied to any draughting medium. Pressure is then applied to the character using a burnisher, ballpoint pen or similar tool, taking care to rub over only the character required. The character can be seen when it is detached from the carrier film; the carrier is then carefully peeled back, leaving the character adhered to the drawing sheet. The backing sheet is now placed over the character and reburnished to give maximum adhesion.

When transfers have been used on tracing paper or film it is advisable to apply a fixative spray from an aerosol. The fixative prevents damage to the transfer which might occur during the printing process or in storage.

Any errors which occur during application of transfers can be rectified using a razor, an eraser, or by lifting the transfer with adhesive tape.

Transfers provide the draughtman with a quick and efficient method of producing high quality incidentals on a drawing, However, the cost of transfer sheets tend to limit their use to presentation type drawings.

Figure 1.19 is representative of some of the characters available, but for a full appreciation of all the various types, comprehensive catalogues are available from manufacturers.

Recent developments in dry transfer graphics include transfer sheets depicting proprietory building components such as door and window joinery, mechanical services equipment, etc.

These sheets are produced in co-operation with manufacturers to illustrate a product range. Such sheets represent a further

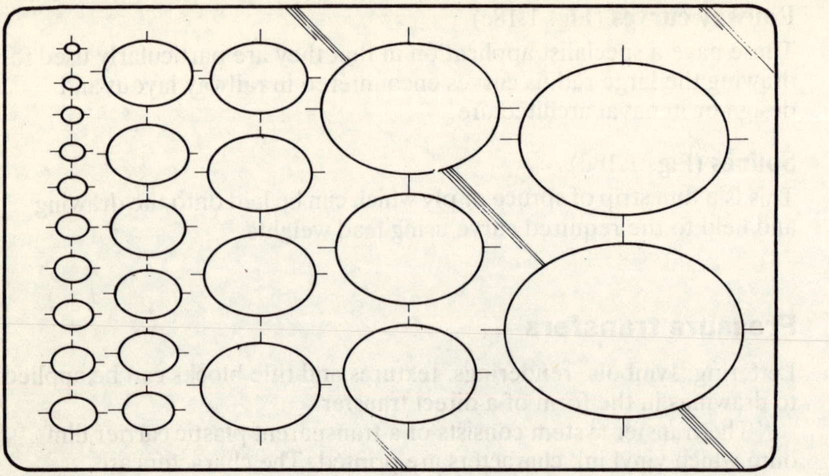

(a) Ellipse aid

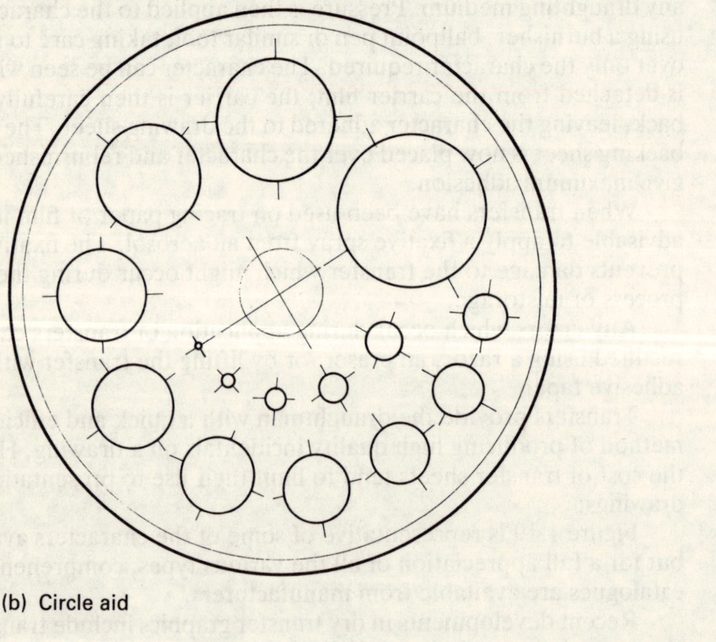

(b) Circle aid

Fig. 1.17

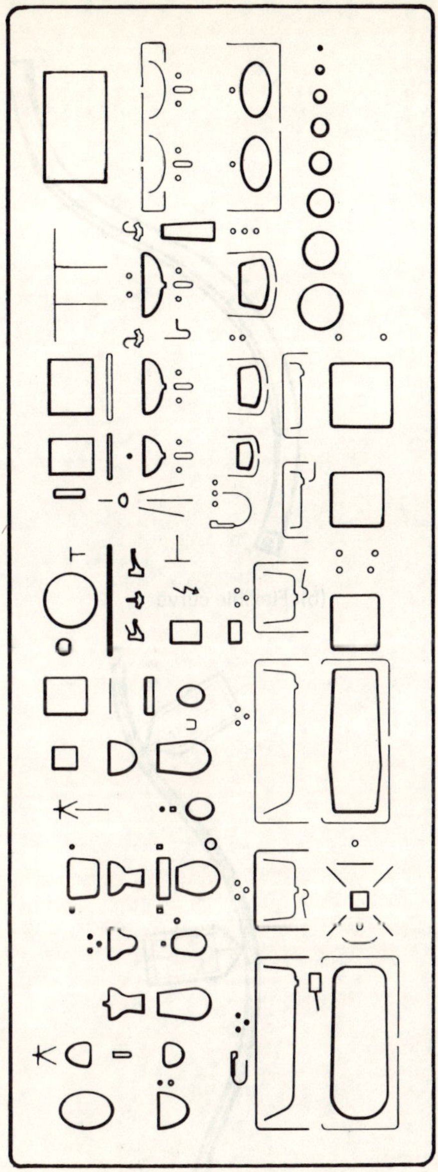

(c) Sanitary ware

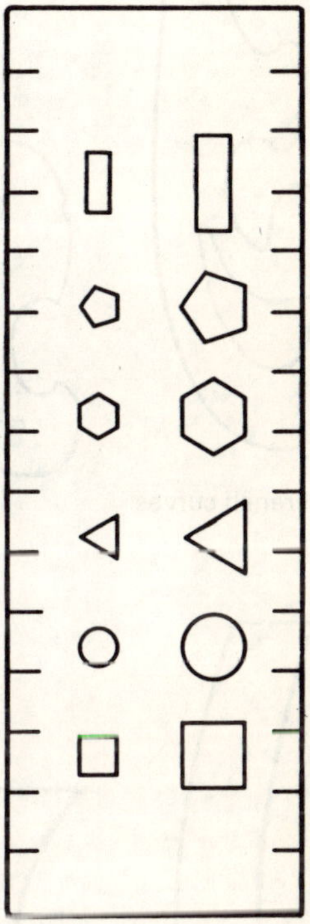

(d) General template

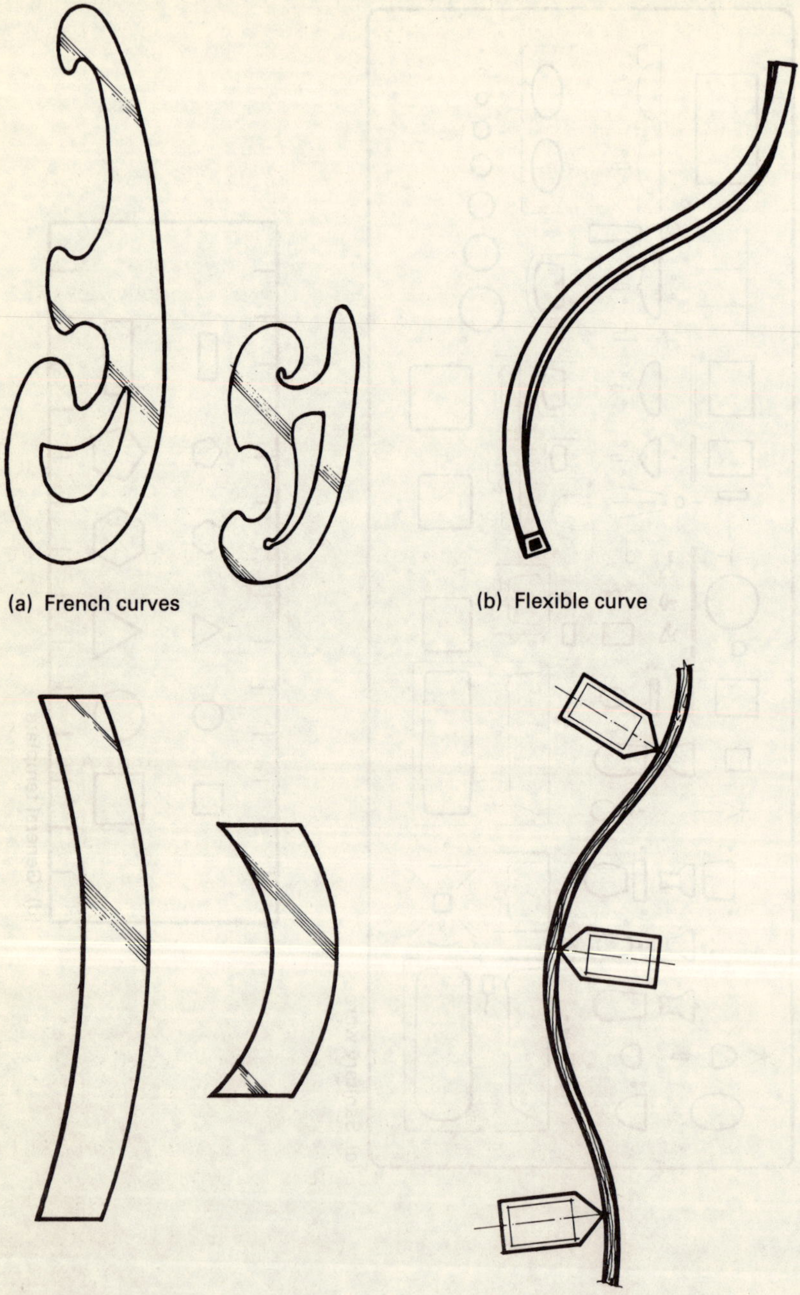

(a) French curves

(b) Flexible curve

(c) Railway curves

(d) Spline and weights

Fig. 1.18

rationalization of the drawing process, and are particularly useful on large projects where considerable repetitive drawing can be avoided.

Adhesive films

Figure 1.20 shows some of the many types of adhesive films which are available for use on tracing paper, to indicate hatching, shading, rendering, or to show various materials and features.

The sheets are printed in a similar manner to that described for pressure transfers, but the films differ in that they have a self-adhesive back which is protected by a backing sheet.

The application of film on a drawing must be carefully executed to avoid damage to the drawing sheet. The usual method of application is to place the backed film over the area to be treated; using a sharp knife, cut an oversized piece of film, taking care not to slice through the backing. The film is then peeled away from the backing and placed on the drawing. The surplus film is trimmed away with a sharp knife to the required outline, and finally burnished to give maximum adhesion.

The effect of films is particularly pleasing on final printed drawings, but the relatively expensive nature of the material does tend to restrict its use to presentation or graphic art work.

Folding of drawings

By standardizing paper sizes and title blocks it is also possible to approach the folding of drawings in a similar way.

The student will initially be executing drawings on cartridge paper, probably on A2 sheets, and to enable a neat folio of drawings to be kept consideration should be given to accurate folding. In practice folding techniques are equally important, for example, when preparing dye-line prints for dispatch. It must be made clear however that masters drawn on tracing paper are never folded.

Folding for clip binder storage

As previously mentioned, a 20 mm border should be provided on the left-hand side of the drawing sheet to permit holes to be punched without securing all the folds of the drawing sheet.

The sequence for folding an A2 sheet is shown in Fig. 1.21a and it should be noted that when complete the title block appears on the front leaf, so that the drawing can be quickly identified without the necessity of opening out the sheet. With a little practice it is possible to fold accurately without the need to measure and, indeed, some preprinted tracing sheets have fold marks printed on them.

Further methods of folding for various sheet sizes can be found in BS 1192.

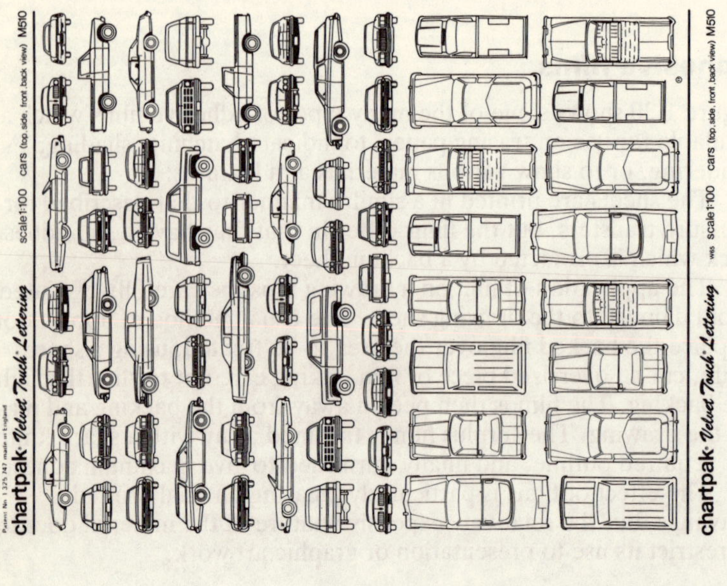
Patent No 1 325 747 made in England
chartpak Velvet Touch Lettering
scale 1:100 cars (top, side, front, back view) M510
chartpak Velvet Touch Lettering
scale 1:100 cars (top, side, front, back view) M510

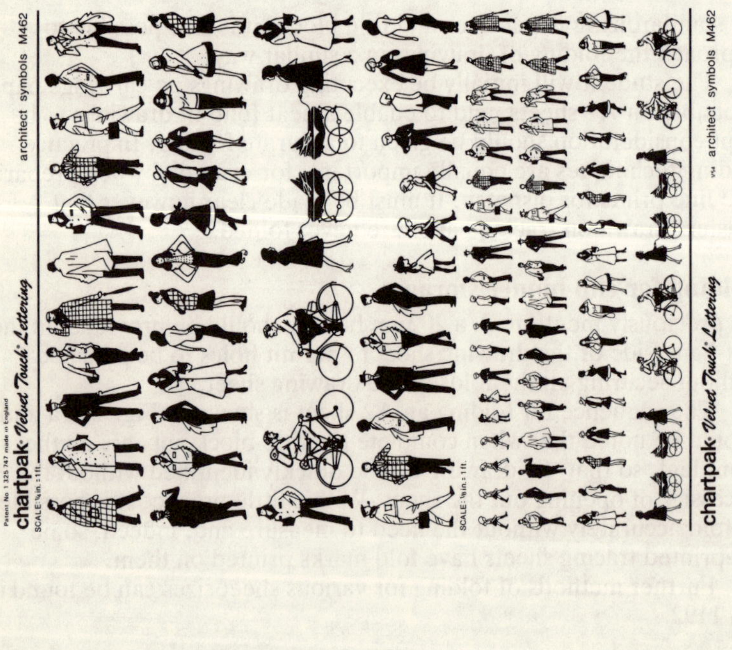
Patent No 1 325 747 made in England
chartpak Velvet Touch Lettering
SCALE 1/8 in. = 1 ft.
architect symbols M462
chartpak Velvet Touch Lettering
architect symbols M462

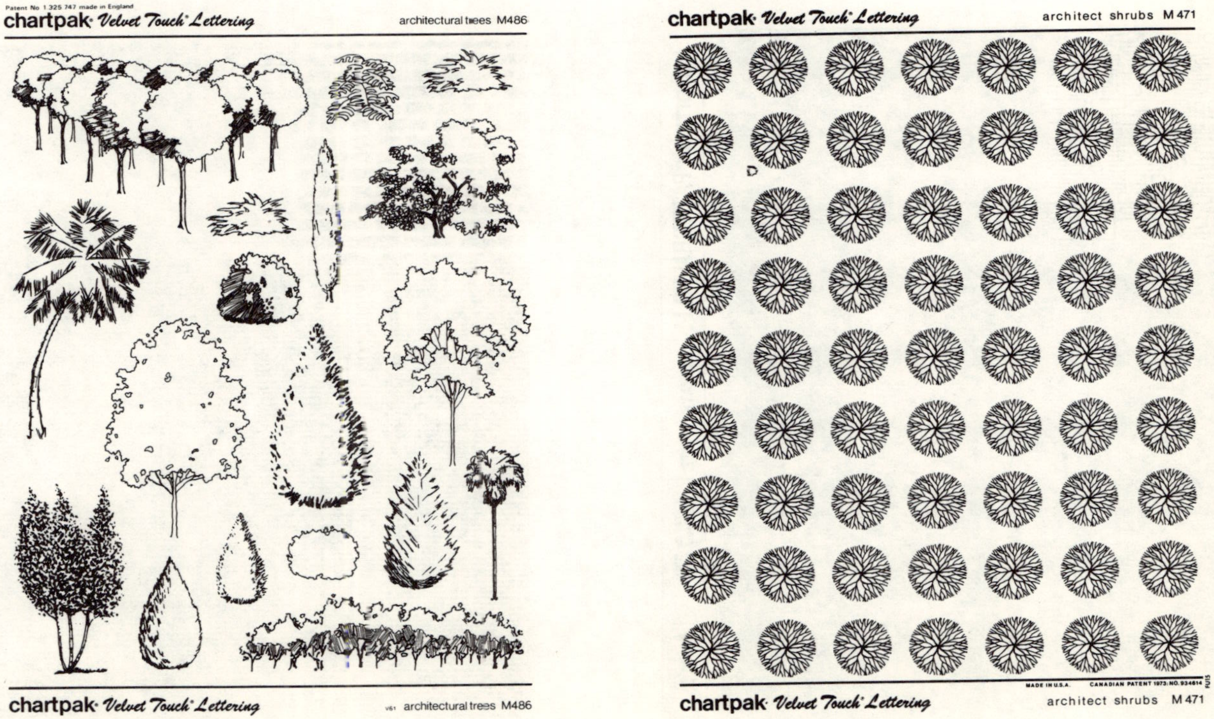

Fig. 1.19 Pressure transfers

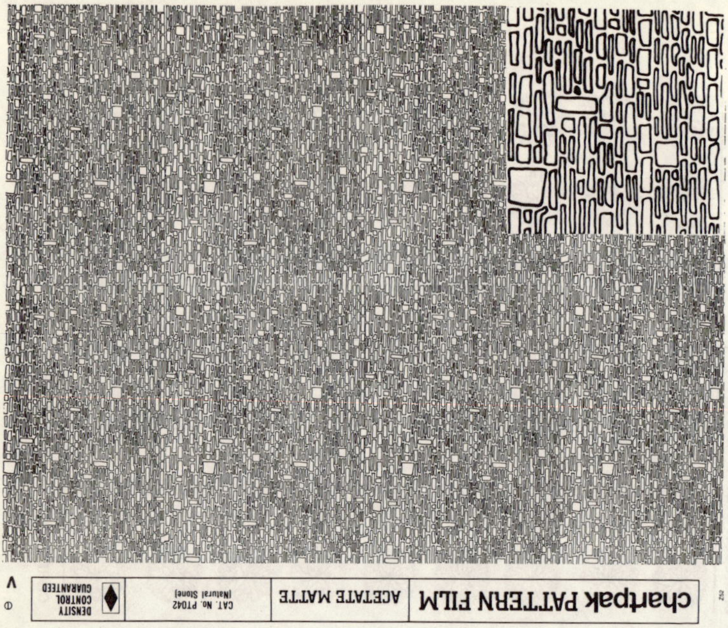
chartpak PATTERN FILM
ACETATE MATTE
CAT. No. PT042
(Natural Stone)
DENSITY
CONTROL
GUARANTEED

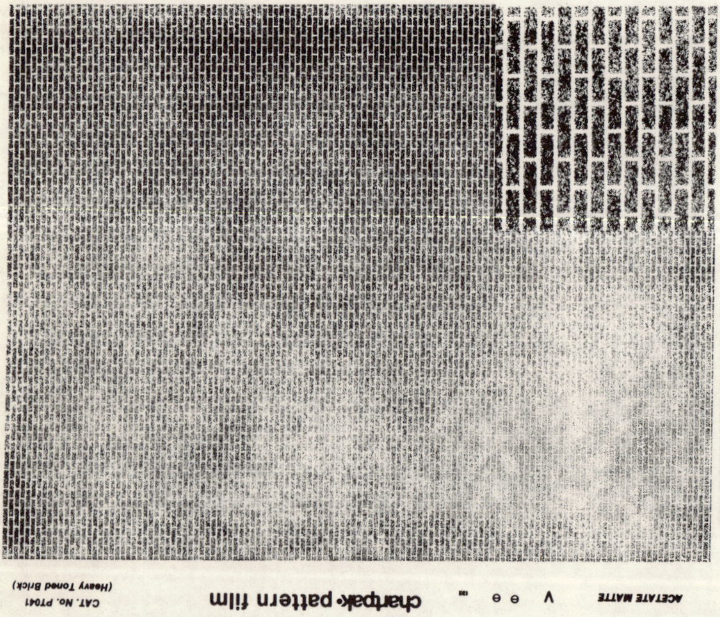
ACETATE MATTE
chartpak• pattern film
CAT. No. PT041
(Heavy Toned Brick)

Fig. 1.20 Adhesive films (insets show exact size)

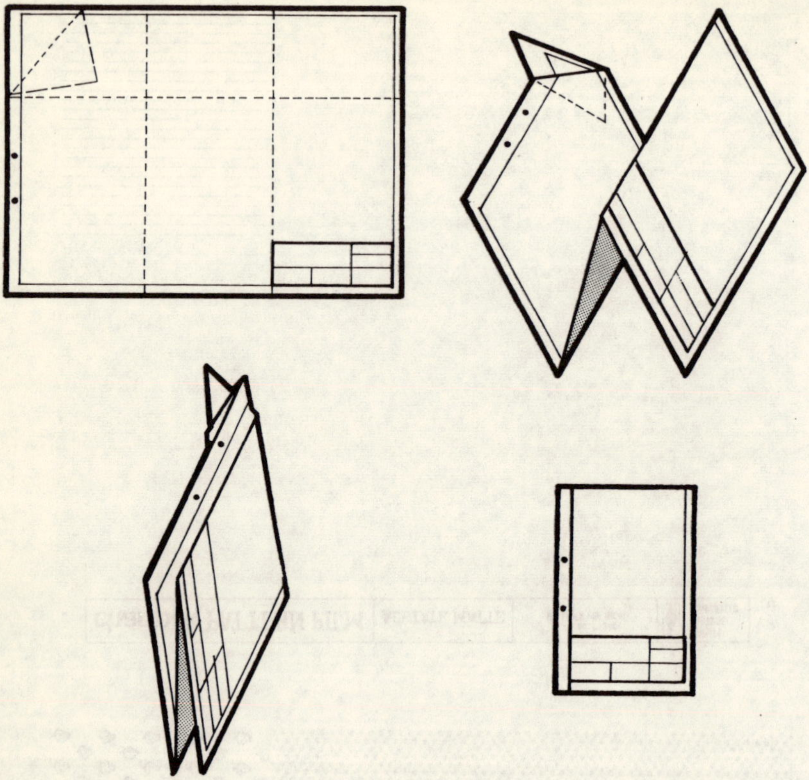

Fig. 1.21(a) Folding drawings

Folding without binding edge

The sequence for folding an A2 sheet without binding edge is shown in Fig. 1.21b, and needs no further explanation.

Consistency in approach to folding drawings, although it may seem relatively unimportant, can save considerable time and frustration which so often arises when trying to re-fold a drawing particularly on site.

Reprographic processes

Reprography can be considered as all the document reproduction processes carried out in an office on a non-commercial scale. It follows that reprography is in itself an extensive subject and accordingly this

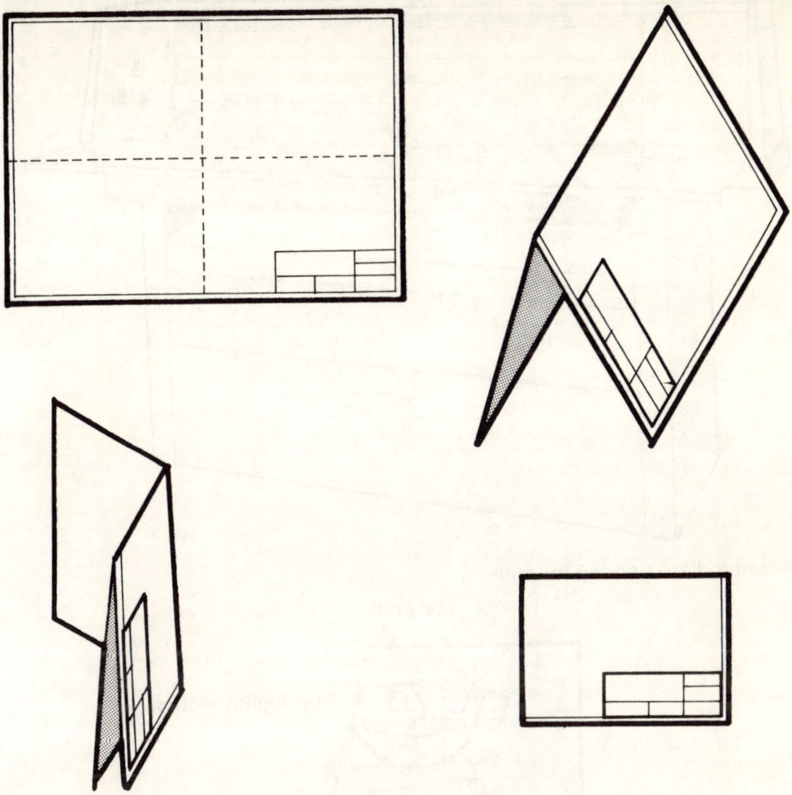

Fig. 1.21(b)

section will deal only with processes in current use for the reproduction of drawings.

Dye-line (sometimes known as 'diazo' or 'white printing')

This method is probably the most extensively used largely due to its simplicity, economy and also the fact that the process will accommodate the very large prints often required in drawing offices.

The process is based upon a translucent master through which light can pass, and a light-sensitive copy paper onto which the master image is transferred.

The translucent master is placed face up onto a sheet of copy paper which is coated in diazonium salts. The combination is passed through a machine which exposes it to a high intensity ultra-violet light source.

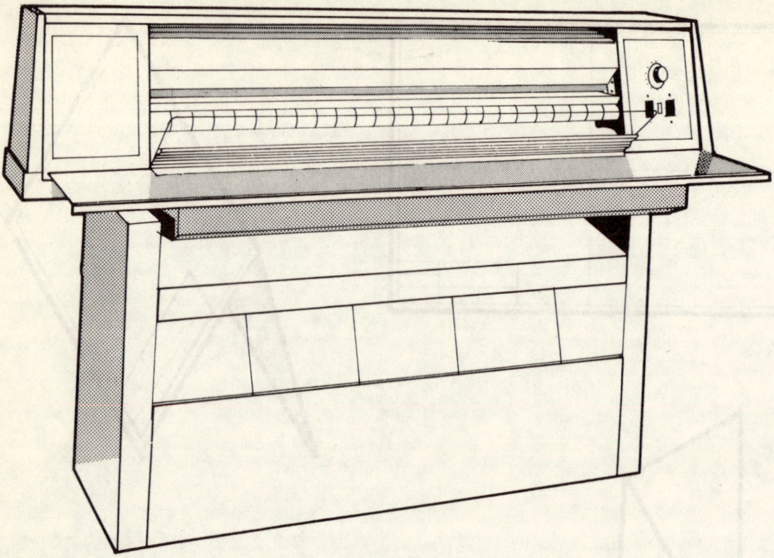

Pressure diazo printing machine

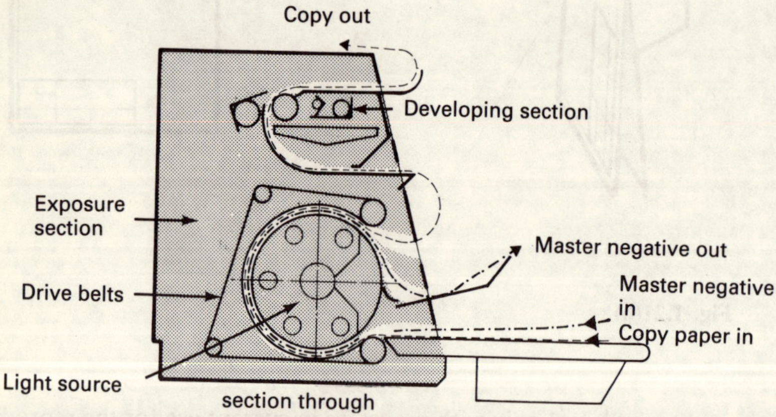

Fig. 1.22 (Source: AM Admel, a division of AM Bruning Ltd.)

The effect of the light is to destroy the salts so exposed, but to leave the salts unaffected in the unexposed areas which are shielded from the light by the linework of the drawing.

Having been exposed the master and copy paper are separated, and the copy paper is put through a development process which renders the salt remaining on the paper clearly visible. The process of development varies with differing machines, but generally those in common use are the semi-dry process and the ammonia process.

The semi-dry process

In this developing process the copy paper is passed over a roller which rotates in a trough of developing liquid and in turn moistens the copy paper, hence the term semi-dry. The effect of the fluid is to produce and fix a visible image on the copy paper. The liquid used in the process is volatile, so care must be exercised when in use. The machine has to be regularly cleaned to prevent corrosion from the developing fluid and to avoid the build up of crystalline deposits, a further characteristic of the fluid. The output of this type of machine is less than its ammonia counterparts and the print quality is slightly inferior. It does, however, have the advantage that ducting and ventilation is unnecessary, unlike ammonia machines.

The ammonia process

The ammonia developing system produces a dye image by the action of ammonia gas on the light-sensitive areas of the copy paper. Ammonia is fed into the machine in liquid form, and is vapourized to gas by the application of heat. This process is quicker than the semi-dry process, having a feed rate of over 20 metres per minute, but, at the same time, producing a superior quality print to the semi-dry process. Its principal disadvantage is that excess ammonia gas must be ducted away from the print room to external air, as the fumes are unpleasant and possibly harmful.

It must be appreciated that the copy paper used in dye line printing is sensitive to daylight. It is important to ensure that the materials, when in use, should be shielded from daylight as much as possible. Failure to do this could prove to be an expensive mistake.

Although when developed a print is to all intents and purposes 'fixed', copies made by the dye-line process do have a tendency to fade over a long period of time, particularly if left in sunlight. It should also be realized that the paper used for copying is not dimensionally stable, particularly when the semi-dry process is used. Scaling from prints is not therefore recommended practice.

For the best results from dye-line printing, master drawings should be carried out in ink. However, with the variation in copy paper speeds and the speed control which can be exercised over the machine, acceptable prints may be obtained from pencil masters. Whichever method and materials are used, the important underlying factor is that the master must be executed on a translucent medium before this method of reprography can be considered. It is not possible to obtain dye-line prints from drawings made on cartridge paper.

Electrostatic copying

This method of reprography, more familiarly referred to as xerography (which is not a trade name when used with a small x), is probably the most extensive method of document copying currently in use, although

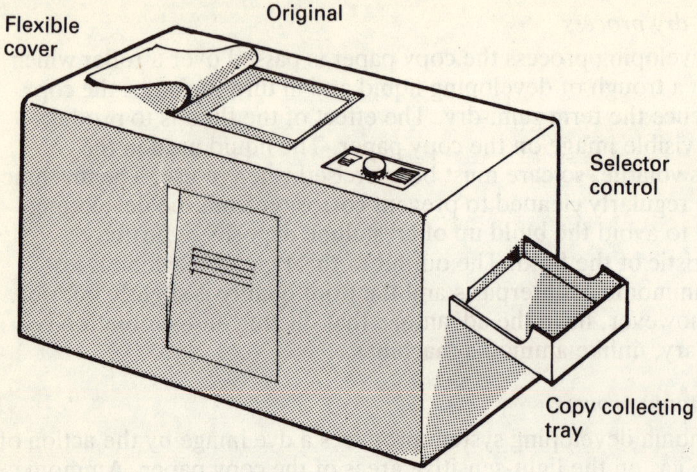

(a) Small electrostatic copier

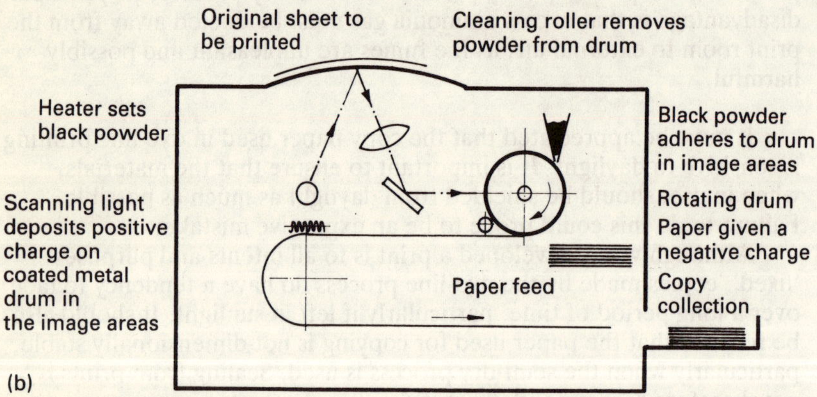

(b)

Fig. 1.23

it is limited generally to a maximum paper size of A3, which does restrict its application to the copying of drawings.

In this type of copier a coated drum is given a positive electromagnetic charge as it revolves, and an image of the original is projected onto it through a lens. In the unprotected areas of the drum the charge is dissipated by the light, but is retained in the image areas, shielded from the light. As the drum continues to revolve it is cascaded with developer powder which adheres to the charged image areas. The powder retained on the image areas of the drum is then transferred to the negatively charged copy paper as it passes around the revolving drum. The powder now deposited on the paper is fixed by a heater unit, the surplus powder being cleaned from the drum in preparation for the next cycle.

The process is quick, dry, permanent and the quality of reproduction is generally very good. Documents on opaque paper as well as tracing paper masters can be reproduced, but because of the size limitation it is often used in conjunction with a dye-line machine.

Figure 1.23b shows a simplified section of a typical electrostatic machine.

Offset lithography

Although this particular process has a limited application for the reproduction of working drawings, the use of miniature offset litho is becoming more and more commonplace for general document reprography, where large numbers of copies are required. It is indeed the capacity of offset litho to produce large numbers of good quality copies comparatively cheaply which makes the process attractive.

The method of copying is based upon ink transfer from a master plate onto copy paper via a system of rollers.

The master plates may be prepared on paper, plasticized paper, plastic or metal, the choice of medium being related to the number of copies required, i.e.

(a) Paper plate, up to 1000 copies.
(b) Plastic plate, up to 10 000 copies
(c) Metal plate, up to 50 000 copies

Masters may be prepared directly onto the paper and plastic plates using special pens or typewriter ribbons, by a reprographic process not dissimilar to electrostatic copying already described, or by photographic methods in which case the plate is given sensitized coating. Plates prepared using photographic techniques have unlimited use.

The process is shown in diagrammatic form in Fig. 1.24. The plate is attached to the plate cylinder (A), which revolves. In doing so the master plate is damped by the damping roller (B), which damps the plate. The image area, being greasy, does not however absorb the moisture.

The inking roller (C) inks the image areas but the ink is not accepted by the dampened surfaces.

The blanket cylinder (D) accepts the ink from the master plate and transfers the image from the master to the copy paper as it is fed over the impression roller (E)

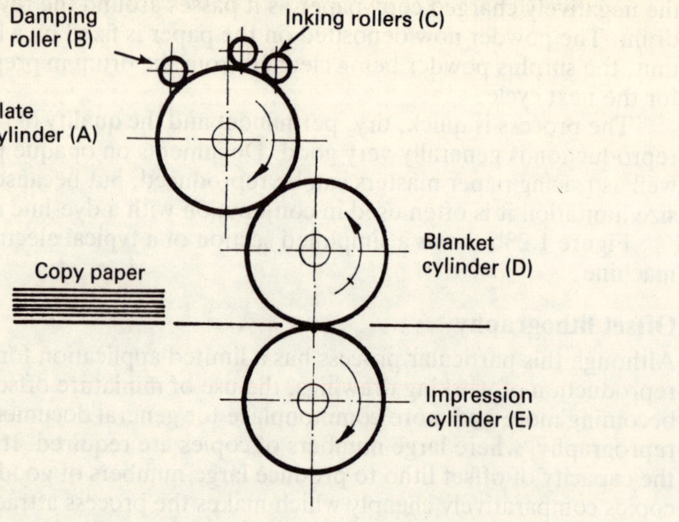

Diagramatic section of litho copy process

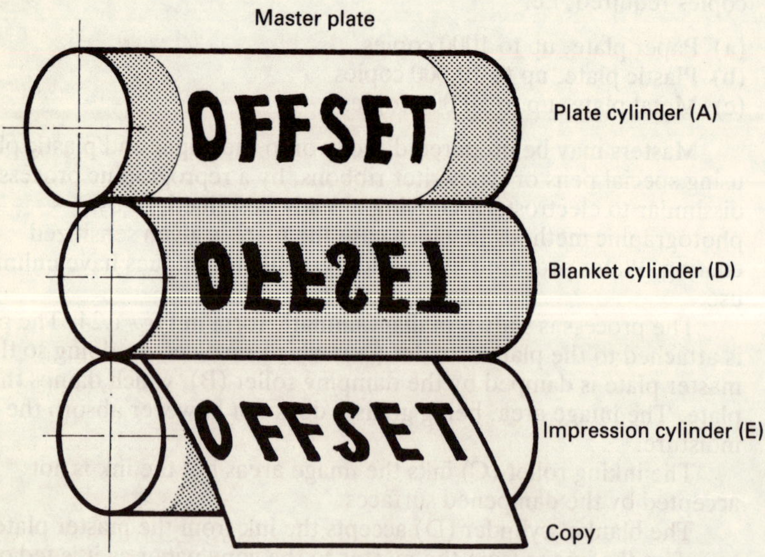

Image reversal

Fig. 1.24

Chapter 2

Drawing Procedures

Setting up

The drawing sheet should be securely fixed to the drawing board using adhesive drafting tape, or preparatory board clips. Drawing pins should not be used as these damage the drawing board and tend to increase the likelihood of the drawing sheet being torn.

The sheet should be secured to the board such that the horizontal edges of the sheet align with the datum edge of the tee square; this ensures that the drawing will be square relative to the edges of the sheet.

It is recognized practice to draw a border line around the edge of the sheet, not only to effect some safeguard to the drawing should the edge of the sheet become damaged, but also to permit filing tags or binder holes to be accommodated without obscuring any of the linework. In general the border is drawn 10 mm from the edge of the sheet. If, however, it is intended to make provisions for filing or binding, a 20 mm border is drawn at the left-hand end of the sheet, as shown in Fig. 2.1a.

To assist in consistent setting up it is useful to lightly draw a horizontal datum line (loosely termed the x, y line) across the sheet, so that if the drawing has to be realigned, having been removed from the board, it can be done accurately by reference to the datum.

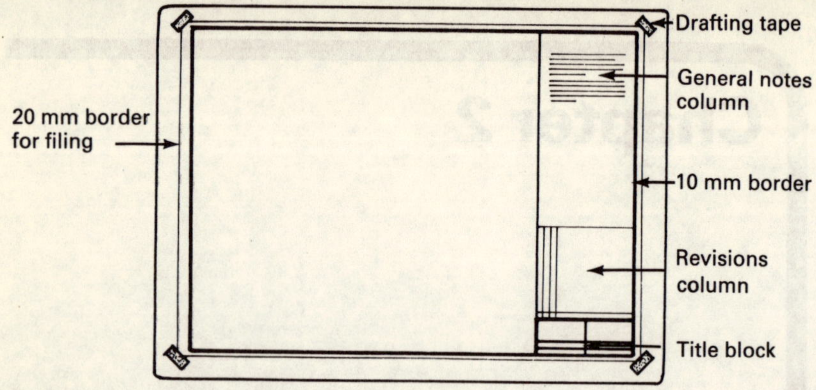

(a) Preparation of drawing sheet

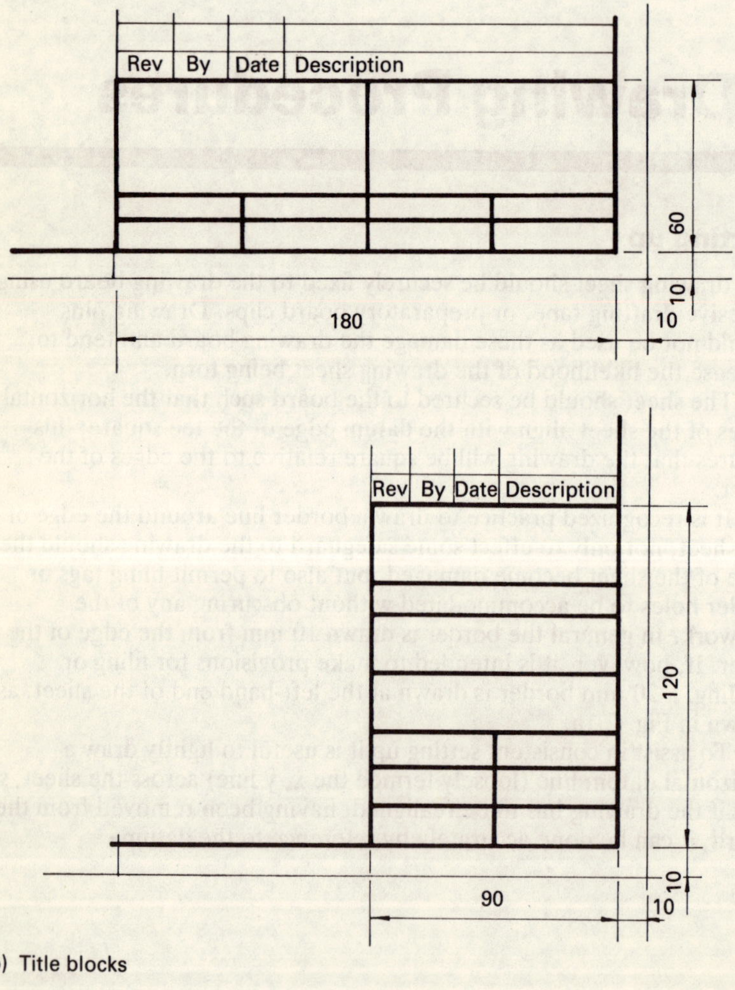

(b) Title blocks

Fig. 2.1

Title blocks

Although in practice considerable variation occurs in the presentation of titling, BS 1192 does contain recommendations regarding sizes, position and layout of title blocks. The standardization of title blocks in this way ensures that, when the drawing is folded in the correct manner, the description of the drawing shown upon the title block will be clearly visible without the necessity of opening out the drawing.

The title block should be located in the bottom right-hand corner of the sheet and should convey the following information:

1. A concise description of the work.
2. The scale(s) used.
3. The date of production.
4. The originator of the drawing
5. The drawing number
6. Revisions made to the drawing.

The recommended format for title blocks is shown in Fig. 2.1b; the content of the block can be adapted to suit the student's drawing.

The wording within the title block should be brief and concise, and executed in bold lettering to make it conspicuous.

Manufacturers of graphic transfers produce title blocks for use by professional practices, and indeed it is possible to obtain preprinted tracing sheets having border, title block and standardized notes printed directly onto the sheet. Although relatively expensive, they represent a considerable time saving in the drawing office, as well as ensuring standardization in presentation.

Linework

The mastery of drawing in pencil on cartridge paper provides the student with an excellent introduction to draughting, and in common with many pursuits the essential element in acquiring a high degree of skill is practice.

The student should firstly realize that in drawing a single straight line the following considerations should be made:

1. Quality of line
2. Weight of line
3. Continuity
4. Value.

Line quality

A drawing is an ordered arrangement of lines and it is obvious therefore that line quality is an essential ingredient in the production of acceptable drawing presentation.

Pencil preparation and quality has already been mentioned and needs no further explanation, other than to say that if these simple preliminaries are not carried out it is not possible to produce quality linework.

The aim of the student is to produce a line which is:

1. **Firm**. The line should be crisp and clear and not take on a 'furry' or double appearance. Finished lines must have crisp ends and not fade indeterminately.
2. **Straight.** The pencil point must not be allowed to wander away from the straight edge as this will result in a wavy line. The stock of the tee square must be held firmly against the edge of the drawing board.
3. **Consistent.** The line should not vary in thickness or density throughout its length, neither should there be unintentional breaks within its length.
4. **Clean.** Precautions already described will assist in keeping lines clean, but as a general rule it may help to work from top to bottom and from left to right of the drawing sheet. This will reduce potential smearing, as the squares will not be required to travel over the completed linework so frequently. Left-handed people should of course work from right to left.

Weight

The weight of a line can be varied on a drawing to improve clarity. An example of this might be where a light line is used to indicate brick coursing on building elevations, the variation being achieved by using a harder grade pencil.

Continuity

The relationship of lines is important insofar as the clarity of a drawing is concerned. By giving lines of equal importance similar value, interpretation of the drawing is simplified and presentation is accordingly greatly improved. The student should therefore aim to be consistent with the varying types of lines such as outlines, hidden detail lines, section lines, etc., throughout a drawing exercise, bearing in mind that the *value* of each type of line will vary. Various types of line in conformity with BS 1192 are shown in Fig. 2.2a.

Value

Various lines on a drawing have differing degrees of importance. For example, the main outline of an object is more vital than the cross hatching, therefore a thicker line work would be expected for the outline and a fine line for the cross hatching. The use of thick pencil lines on drawings is acceptable provided they fulfil the fundamental requirements for linework.

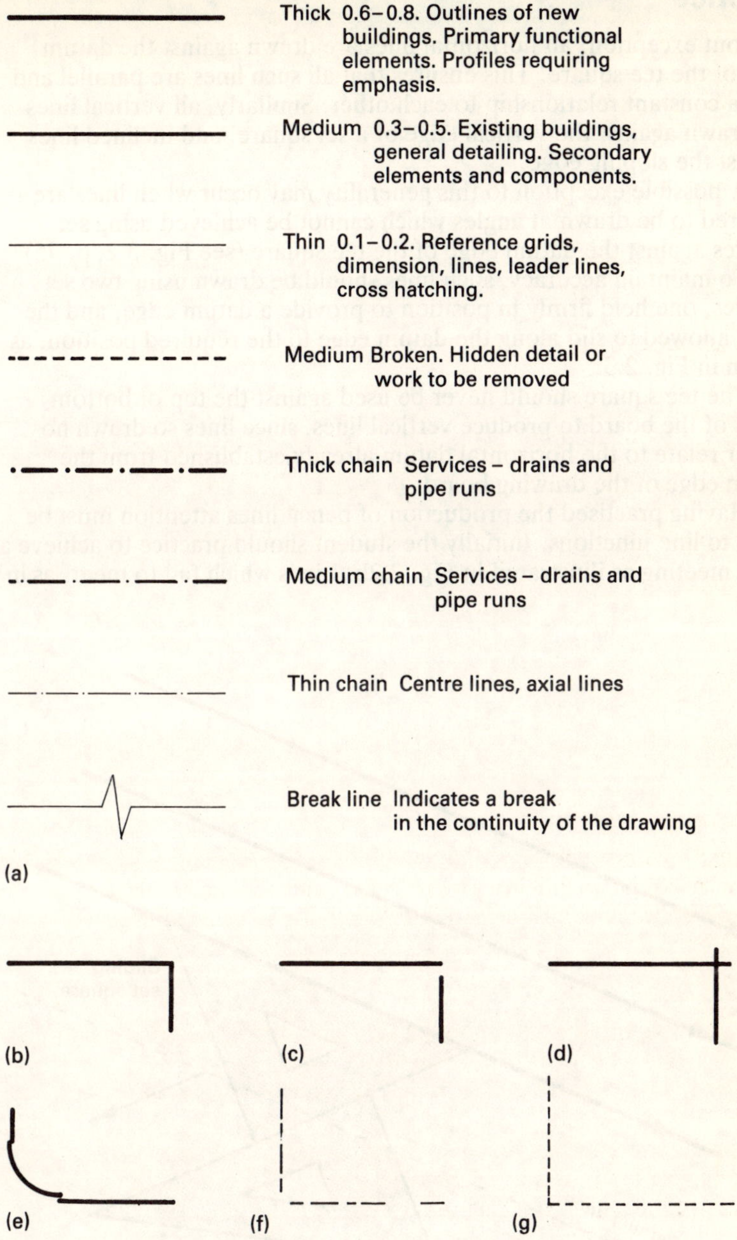

Fig. 2.2 Lines and linework

Practice

Without exception, all horizontal lines are drawn against the datum edge of the tee square. This ensures that all such lines are parallel and bear a constant relationship to each other. Similarly, all vertical lines are drawn against the vertical edge of a set square, and inclined lines against the sloping edge.

A possible exception to this generality may occur when lines are required to be drawn at angles which cannot be achieved using set squares against the datum edge of the tee square (see Fig. 3.2, p. 75).

To maintain accuracy, such lines should be drawn using two set squares, one held firmly in position to provide a datum edge, and the other allowed to slid along the datum edge to the required position, as shown in Fig. 2.3.

The tee square should never be used against the top or bottom edges of the board to produce vertical lines, since lines so drawn no longer relate to the horizontal datum already established from the datum edge of the drawing board.

Having practised the production of pencil lines attention must be given to line junctions. Initially the student should practice to achieve a clean meeting as illustrated in Fig. 2.2b. Lines which fail to meet, as in

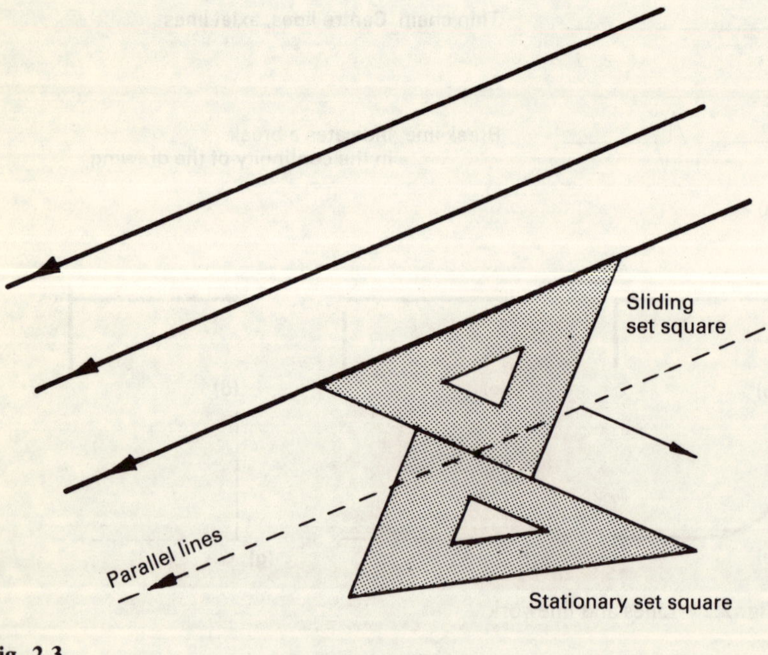

Fig. 2.3

Fig. 2.2c is indicative of a poor draughting standard. It is acceptable in architectural drawing practice to cross lines at junctions as shown in Fig. 2.2d, but without the acquired skill this can look unsightly, so the student should first accomplish the standard illustrated in Fig. 2.2b. To achieve a crisp, well-defined intersection it may prove helpful to reinforce the ends of each line with slightly more pressure on the pencil, working away from the point of intersection.

Points of juncture between arcs and straight lines need special mention. It is essential that the line actually meets the arc, an all too common failing being that indicated in Fig. 2.2e where the arc and line never actually touch. Accuracy is of course all important, but it is generally easier to draw the arc first and then draw the straight line away from the meeting point. A similar method can be adopted where two curves meet, in that the second curve can be drawn away from the meeting point.

Broken lines should be drawn with a consistent line length and correspondingly equidistant spaces between the lines. Junctions between broken lines should be made with a distinct meeting of lines as shown in Fig. 2.2g and not an indistinguishable space shown in the example of a poorly drawn line and junction in Fig. 2.2f.

These fundamental skills should be practised by the student, as the quality of draughtsmanship ultimately achieved will depend largely upon the points of detail described.

Approach to layout

The pleasing appearance of a finished drawing will depend, amongst other factors, upon the layout and relationship of the various views on the drawing sheet.

To avoid an unbalanced presentation due consideration must be given to:

1. The number of views required and their relative positions on the drawing sheet.
2. The scale to be used.
3. The size of the drawing sheet to be used.

If the student devotes adequate thought to these points before commencing the drawing, a balanced drawing which adequately fills the drawing sheet should be achieved. Drawings which are unbalanced with all the views bunched up in one corner, or concentrated to one side of the sheet, will never be pleasing to view.

Having made these preliminary decisions the first steps in any drawing should be to set out the views in 'block' form to the relevant scale. This preliminary line work will form a basis upon which the drawing can be built up, and as such should be carried out in very light

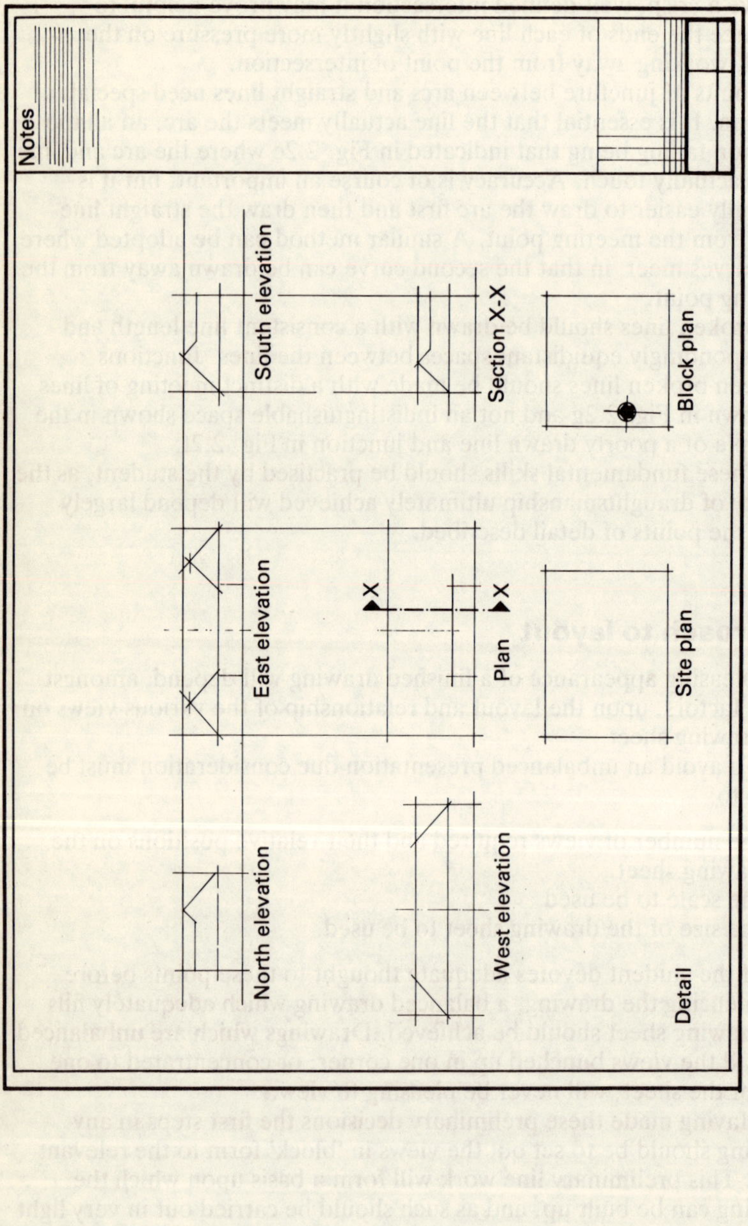

Fig. 2.4 Typical layout of location drawing for a pair of bungalows

pencil lines which should not require erasing. A typical preliminary block layout is shown in Fig. 2.4.

Attention can now be turned to the individual views, adding detail in a very light line, remembering that certain of the lines at this stage will be purely for construction purposes only, and also that the inevitable mistakes made during the building up process can be easily erased.

When the drawing is essentially completed in this form, the feint linework can be drawn over using the correct line value discussed previously, giving emphasis to the outline. The temptation to use bold linework during the early stages of a drawing should be resisted, as errors and alterations cannot be erased cleanly, resulting in an untidy finished drawing.

Dimensioning

The extent to which a drawing is dimensioned will depend upon the nature and function of the drawing in question. For instance, a location drawing may only be required to show principal dimensions such as room sizes, heights, etc., whereas a detailed working drawing such as might be prepared for a steel-framed building would be fully dimensioned to locate every bolt hole and weld.

It is essential that the draughtsman should endeavour to achieve a high degree of skill in dimensioning and initially to develop a clear and uniform approach so as to avoid any doubt about shape, location or features shown on the drawing. To achieve a satisfactory standard of dimensioning the student should pay particular attention to:

1. The selection of dimensions.
2. The technique of dimensioning.

Selection

1. The objective should be to select correct functional dimensions.
2. Careful selection and calculation of dimensions should ensure the practicality of construction or fabrication.
3. Adequate dimensions must be shown on the drawing to eliminate the need for workmen to calculate.
4. Dimensions should be rationalized, and the same dimension should not be repeated on the same drawing.
5. Unnecessary dimensions should be disregarded and omitted from the drawing.

Technique

The following fundamental rules should be followed when dimensioning:

1. Do not use outlines as dimension lines.

2. Place dimension lines outside of outlines wherever possible.
3. Do not cramp dimensions and dimension lines; think ahead and make the necessary allowances if a large number of dimensions are necessary.
4. Dimension lines and projection lines are drawn in a thin full line.
5. Leave a gap between the outline and the projection line.
6. Extend the projection line beyond the dimension line.
7. Figured dimensions should be positioned normal to the dimension line so that they are read from the bottom or from the right-hand side of the drawing.
8. Use a bold decimal point to clearly distinguish metres and millimetres. Centimetres are not generally used in practice.
9. Avoid cramped figures when dimensioning small gaps. It is accepted practice to place the figured dimension outside the gap.
10. Whenever possible lateral dimensions should be shown on plan views.
11. Vertical dimensions should be shown on sectional views.
12. The overall dimensions should be shown furthest from the outline, and the sub-divided dimensions shown between the overall dimension line and the outline of the object.

Figure 2.5(c) illustrates these fundamental techniques.

Types of dimension line

Figure 2.5a–g shows the various types of dimension line generally used on construction drawings, and describes the situation in which each would be used.

Running dimensions (Fig. 2.5a)

It should be noted that running dimensions are used only for survey purposes and this method of dimensioning should not be used for finished drawings.

The use of running dimensions for survey purposes does reduce the risk of error when transposing the survey onto the drawing sheet. It is particularly important when using running dimensions to ensure that the starting point is clearly defined by the circle, and the direction of travel indicated by the arrowheads.

The figured dimensions should be positioned as close as is practically possible to the respective arrowhead.

Open arrowhead (Fig. 2.5b)

This type of dimension is principally used where a system of modular co-ordination has been adopted. It may also be used generally for overall dimensions.

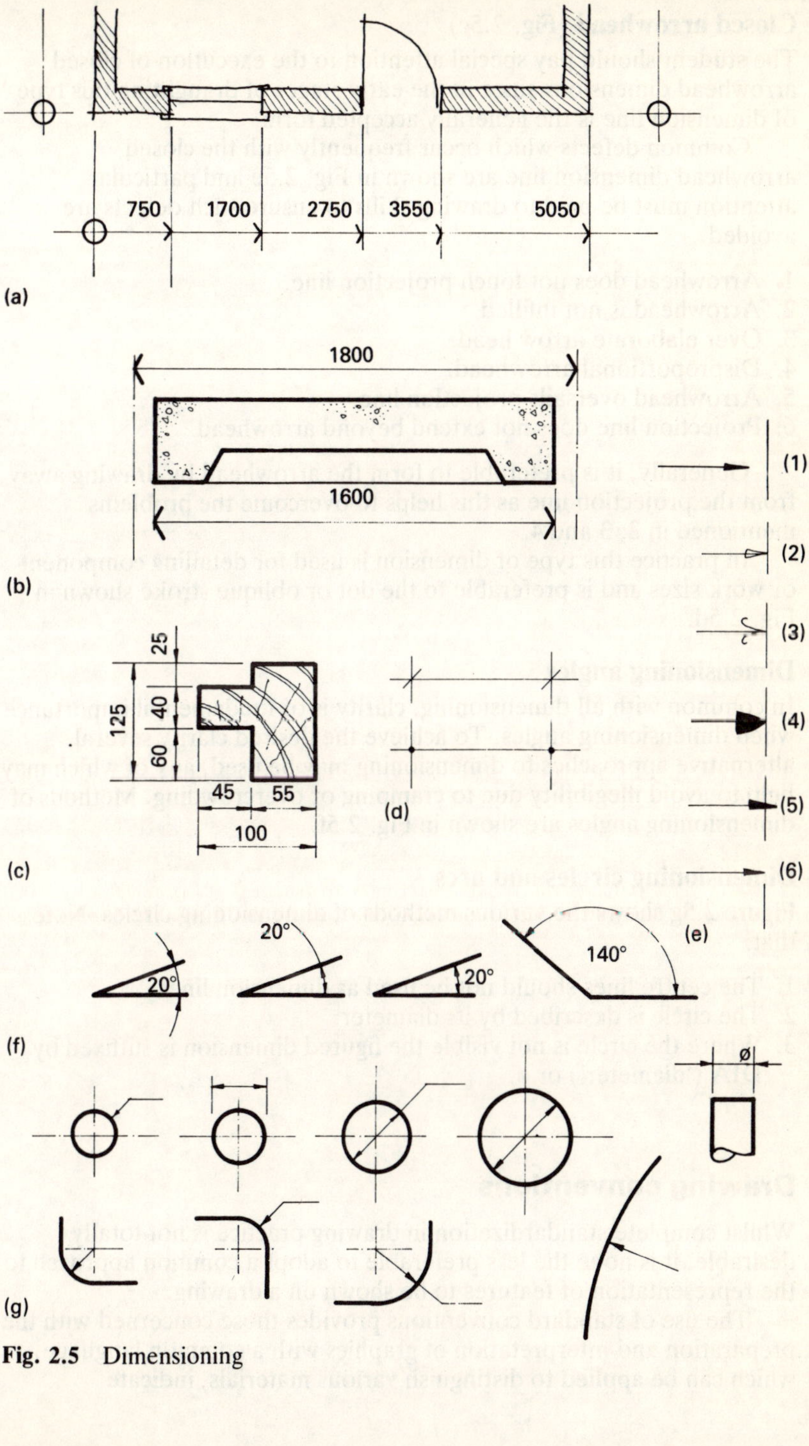

Fig. 2.5 Dimensioning

Closed arrowhead (Fig. 2.5c)

The student should pay special attention to the execution of closed arrowhead dimensions since in the early stages of draughting this type of dimension line is the generally accepted form.

Common defects which occur frequently with the closed arrowhead dimension line are shown in Fig. 2.5e and particular attention must be paid to drawing skills to ensure such defects are avoided.

1. Arrowhead does not touch projection line.
2. Arrowhead is not infilled.
3. Over elaborate arrow head.
4. Disproportional arrowhead.
5. Arrowhead oversails projection line.
6. Projection line does not extend beyond arrowhead.

Generally, it is preferable to form the arrowhead by drawing away from the projection line as this helps to overcome the problems mentioned in 2, 3 and 4.

In practice this type of dimension is used for detailing component or work sizes and is preferable to the dot or oblique stroke shown in Fig. 2.5d.

Dimensioning angles

In common with all dimensioning, clarity is of fundamental importance when dimensioning angles. To achieve the desired clarity several alternative approaches to dimensioning may be used, any of which may help to avoid illegibility due to cramping or overcrowding. Methods of dimensioning angles are shown in Fig. 2.5f.

Dimensioning circles and arcs

Figure 2.5g shows the various methods of dimensioning circles. Note that:

1. The centre lines should not be used as dimension lines.
2. The circle is described by its diameter.
3. Where the circle is not visible the figured dimension is suffixed by DIA ('diameter') or ø.

Drawing conventions

Whilst complete standardization in drawing practice is not totally desirable, it is none the less preferable to adopt a common approach to the representation of features to be shown on a drawing.

The use of standard conventions provides those concerned with the preparation and interpretation of graphics with a common language, which can be applied to distinguish various materials, indicate functions, features, shape, and to clarify information generally.

The relevant British Standards and Codes of Practice should be referred to for the extensive range of conventions, symbols and abbreviations used by the specialist trades within the industry, as only a selected range of those more commonly used can be included in this section.

Reference may be made to:

1. PD 6479: *Recommendations for symbols and other graphic conventions for building production drawings.*
2. BS 308 : *Part 1 1972 Engineering drawing practice.*
3. BS 1192: *1969 Building Drawing Practice.*
4. BS 1553: *Part 1 1977 Specification for graphical symbols for general engineering.*
5. BS 1635: *1970 Graphical symbols and abbreviations for fire protection drawings.*
6. BS 3939: *Section 27 1972 Graphical symbols for electrical power telecommunication and electronics diagrams.*

Figures 2.6 to 2.10 show those conventions most commonly used.

The North point

The North point is used on site plans, block plans and locations drawings to assist in the correct orientation of views. Elevational drawings are described by the principal compass points such that a building elevation which faces South would be described as the South elevation whilst the viewer would be facing North. A selection of typical North points is shown within this volume, all of which are simple to draw and easy to interpret. Flamboyant North points should generally be avoided.

It is good practice when preparing a series of drawings to maintain the same orientation throughout, as this avoids unnecessary problems of re-orientation when each drawing is in use. By reference to the typical layout of a small bungalow shown in Fig. 2.4, p. 48, it will be seen that each elevation is designated by the direction it faces, thus illustrating the importance of a clear North point on the drawing.

Lettering

The art of lettering with a pen is known as calligraphy and is alone a subject worthy of detailed study. This section does not endeavour to cover the topic in detail, but simply to introduce the basic procedures which can be used to achieve an acceptable standard of lettering.

Many students suffer from the frustration of ruining a well-produced drawing by the application of untidy lettering, and few are so gifted that they can produce an acceptable standard of lettering without the need for constant practice. Indeed, there is no mystique

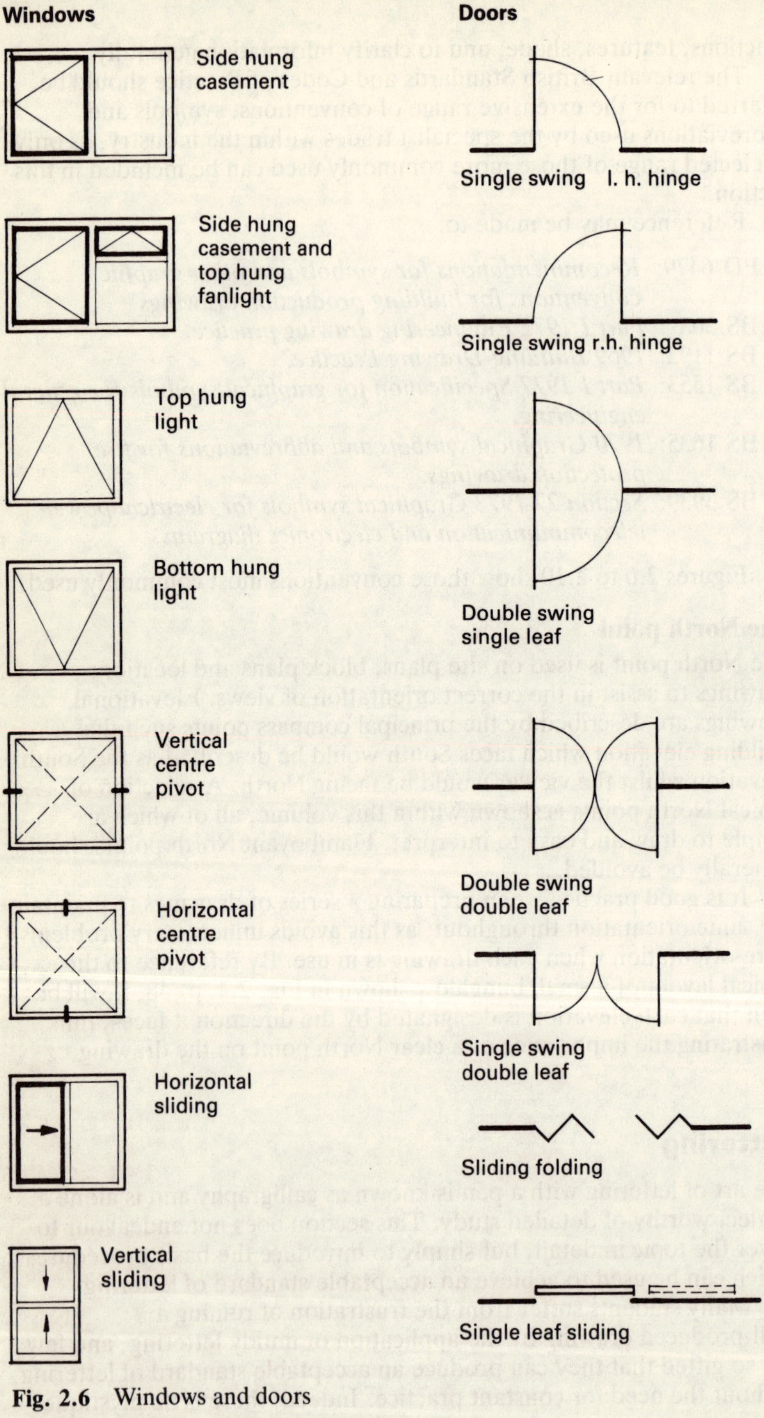

Fig. 2.6 Windows and doors

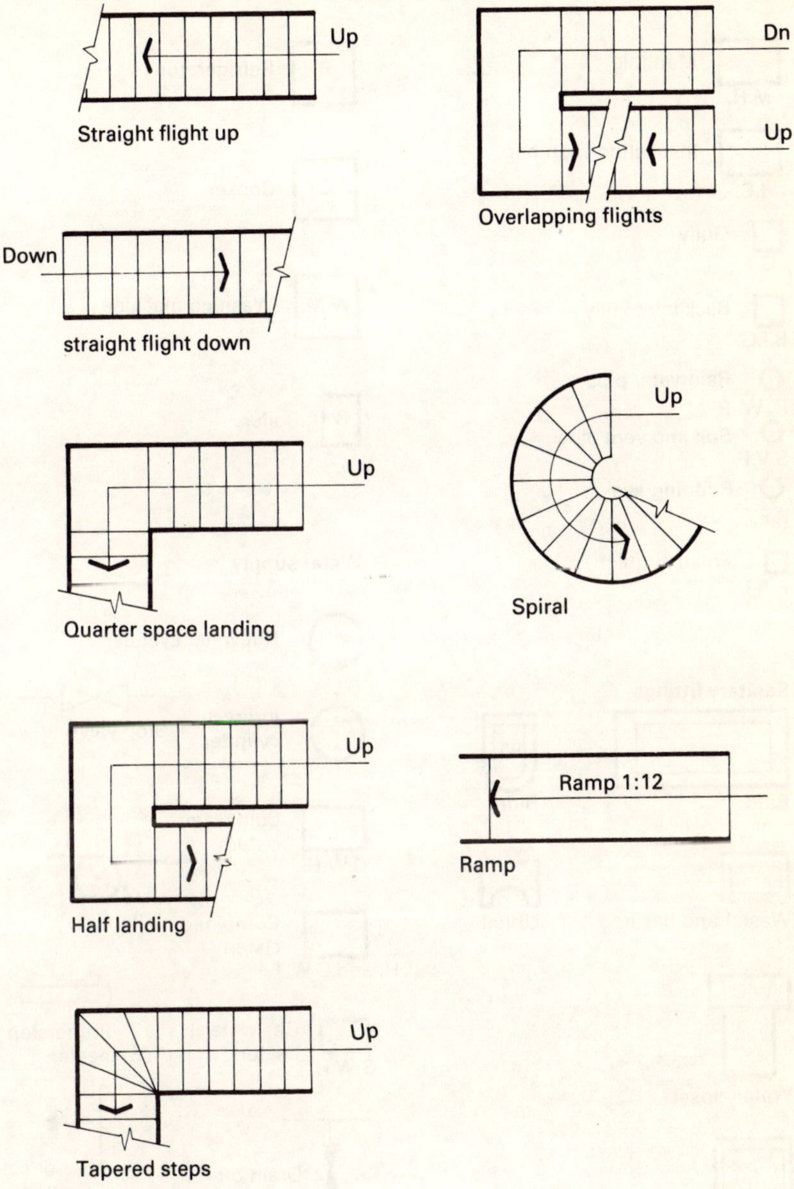

Fig. 2.7 Stairways and ramps

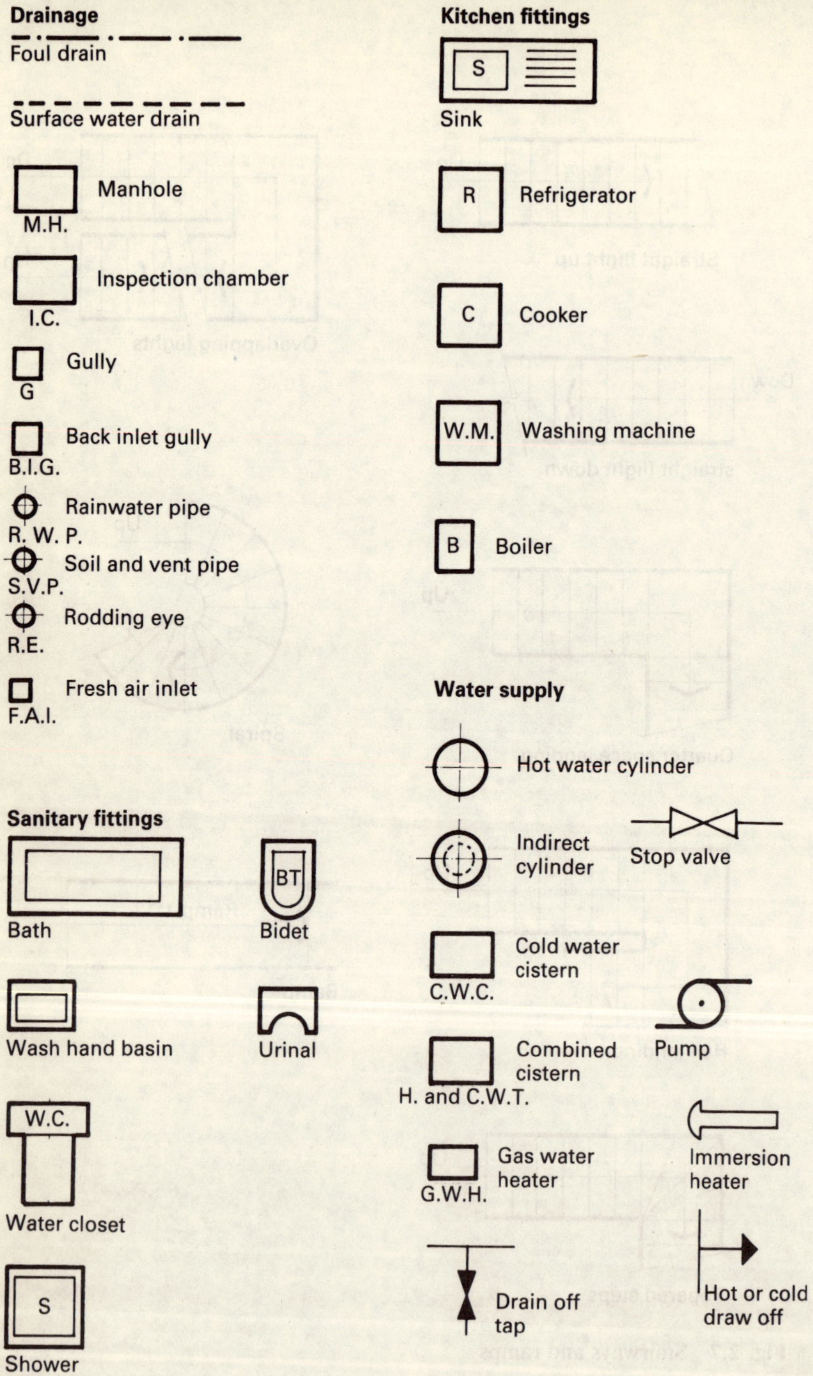

Fig. 2.8 Components

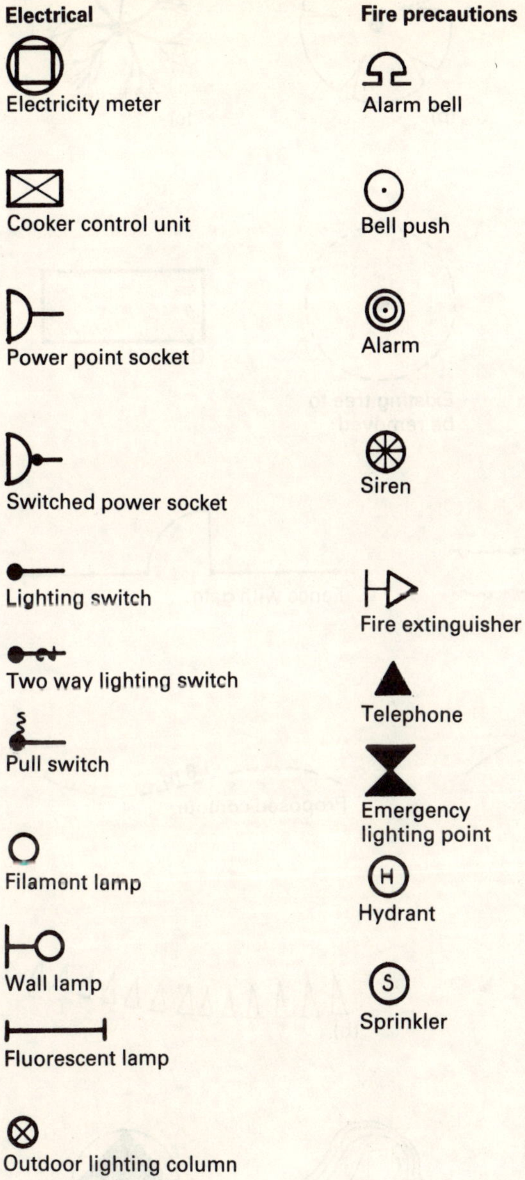

Fig. 2.9 Components

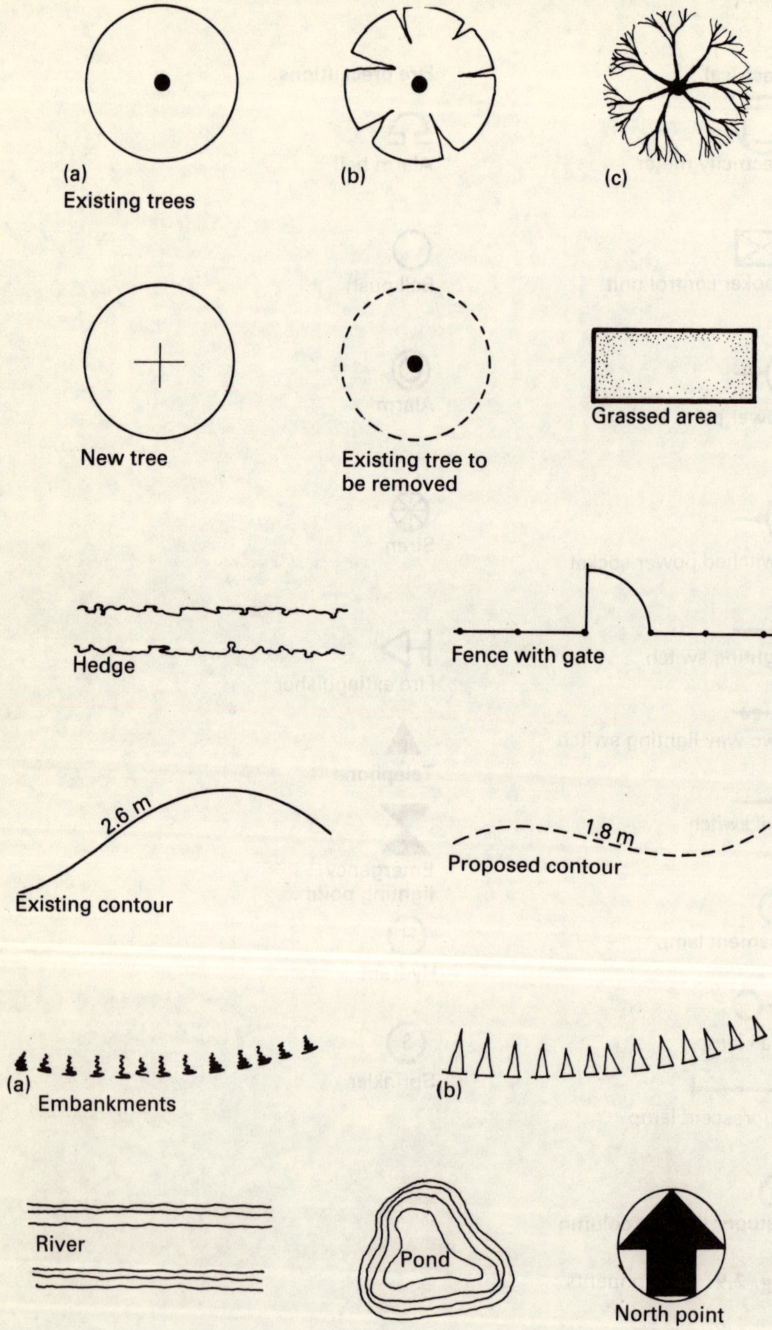

Fig. 2.10 Landscape and survey

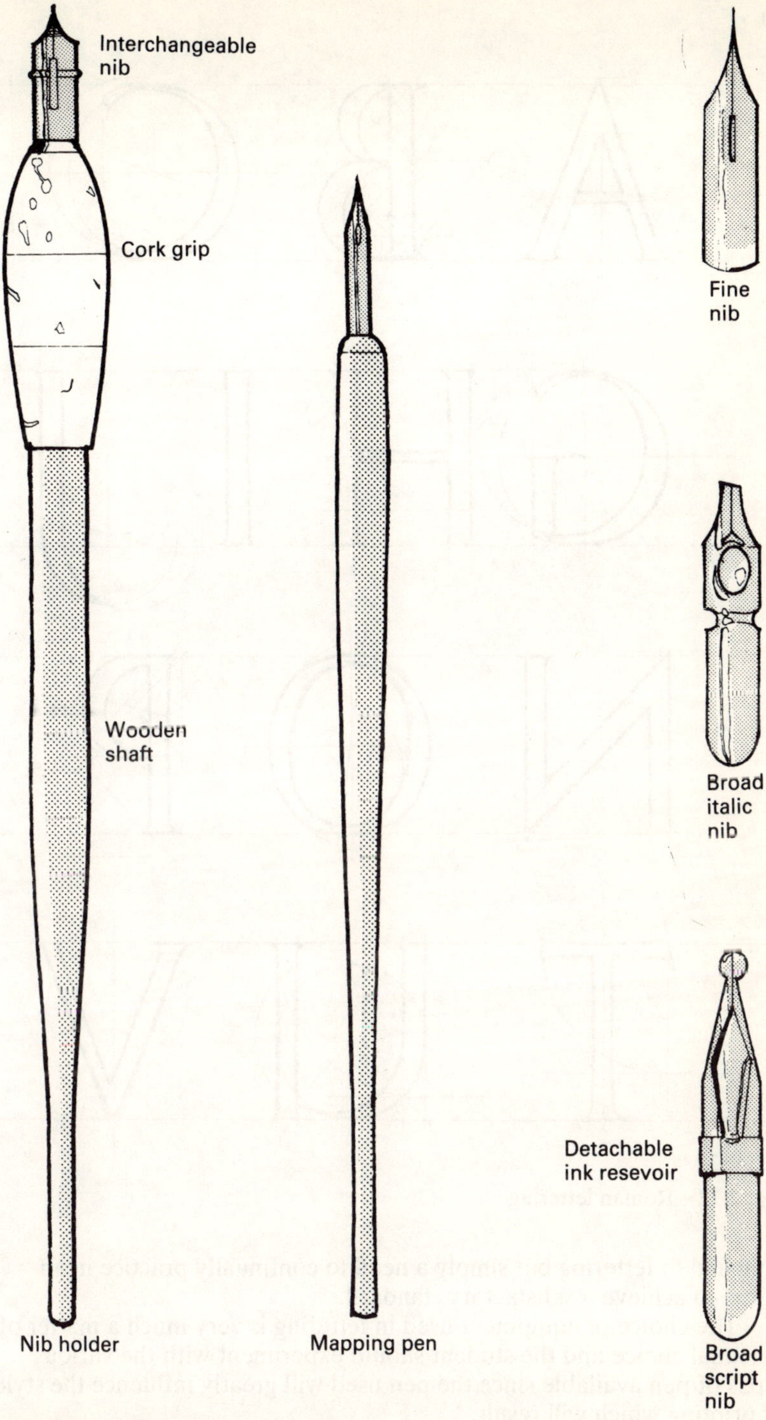

Fig. 2.11 Lettering pens and nibs

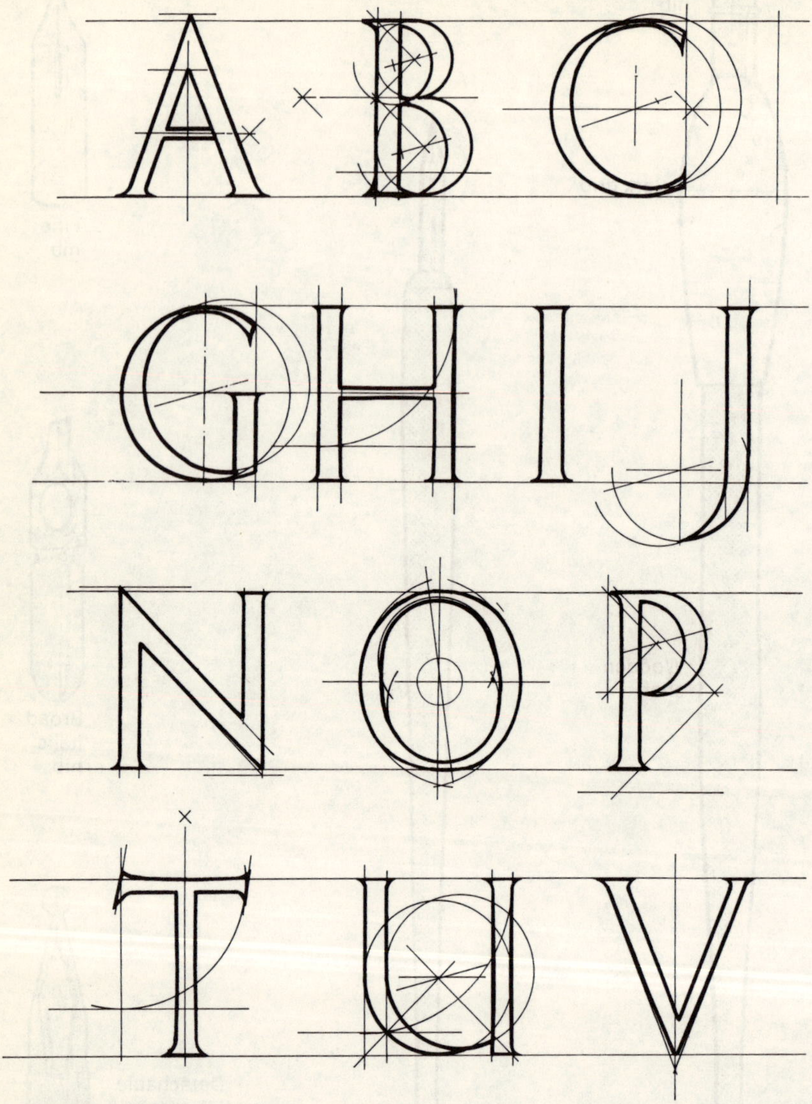

Fig. 2.12 Roman lettering

attached to lettering but simply a need to continually practice in an effort to achieve a satisfactory standard.

The choice of equipment used in lettering is very much a matter of personal choice and the student should experiment with the various types of pen available since the pen used will greatly influence the style of printing which will result.

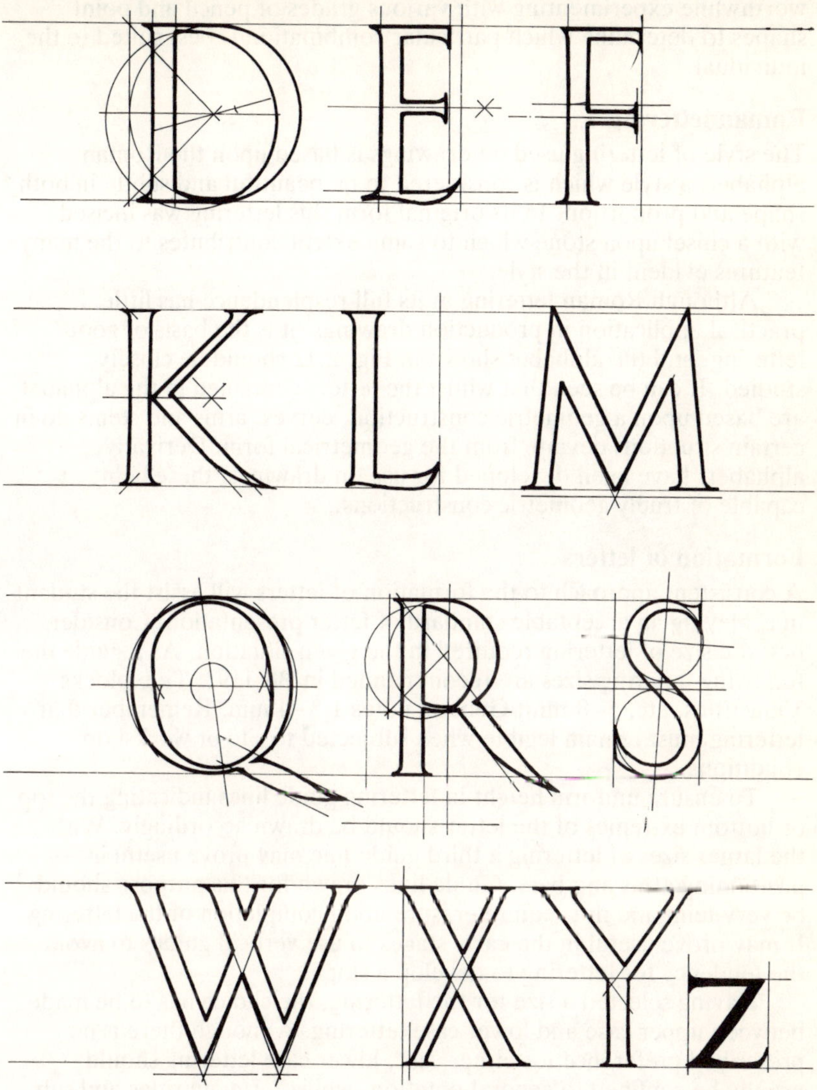

In addition to the type of pen already mentioned in Chapter 1, the student should consider the different lettering nibs used in conjunction with a nib holder. The principal types of nib are illustrated in Fig. 2.11.

Lettering in pencil does to some extent limit the full potential of lettering on a drawing, but none the less it does provide an excellent grounding for acquiring the necessary degrees of skill. Again, it is worthwhile experimenting with various grades of pencil and point shapes to determine which particular combination is best suited to the individual.

Roman lettering

The style of lettering used on drawings is based upon the Roman alphabet, a style which is considered to be beautiful and subtle in both shape and proportion. In its original form this lettering was incised with a chisel upon stone which to some extent contributes to the many features evident in the style.

Although Roman lettering in its full resplendence has little practical application in production drawings, it is the basis of good lettering, and the alphabet shown in Fig. 2.12 should be closely studied. It can be seen that whilst the letters contained in the alphabet are based upon a geometric construction, curves, arms and stems do in certain situations deviate from the geometrical form. Derivative alphabets have been developed for use on drawings, these being capable of truely geometric constructions.

Formation of letters

A consistent approach to the formation of letters will assist the student in achieving an acceptable standard of letter presentation. Consider first the size of lettering required in the given situation. As a guide the following lettering sizes are recommended in BS 1192: Title blocks; View titles, etc. 5–8 mm; General notes 1·5–4 mm. Remember that lettering must remain legible when subjected to site or workshop conditions.

To ensure uniform height in lettering guide lines indicating the top or bottom extremes of the letter should be drawn accordingly. With the larger sizes of lettering a third guide line may prove useful in positioning arms and bars. Guide lines drawn for this purpose should be very feint and not require erasure upon completion of the lettering. It may prove useful in the early stages to use vertical guides to avoid the tendency for lettering to develop a slope.

Having selected a size for the lettering, the choice has to be made between upper case and lower case lettering. Although there is no prescribed preference for either case, lower case lettering should usually be confined to general notation, whilst for main titles and sub titles upper case should be used. Some practices, particularly those with an engineering bias, specify that upper case lettering should be used throughout product drawings.

Numerals

The numerals commonly used on drawings are based upon the arabic style, as opposed to the Roman style numerals (ivx) which are seldom used for drawings. Although based on the arabic style, the numerals are formed in a fashion based upon the roman character and are shown in Fig. 2.13, which shows also the order and direction of pen strokes.

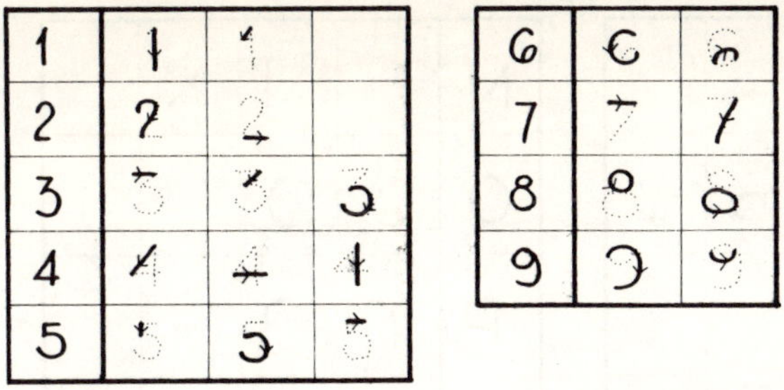

Fig. 2.13

Single line lettering

Single line lettering based upon the Roman alphabet is generally employed on drawings, since with practice this can be executed quite quickly. Legibility is also maintained.

Figure 2.14 shows the single line alphabet and indicates the order and direction of pen strokes. Figure 2.15 shows the lower case alphabet.

Although an individual style will evolve, the use of flourishes should be avoided as they do not contribute to the legibility of the letters. The student should concentrate upon developing skills to produce clear, well-formed letters.

Broad nib lettering

This type of lettering is used with a broad nib such as illustrated in Fig. 2.11. It should be limited to the principal titles and notes since excessive use can become overpowering.

By holding the nib at a constant angle of 45° to the surface, the variation in stroke width will be achieved automatically; firm crisp style should result with practice. A wide range of broad lettering nibs are available, including special nibs for those who are left handed.

Stencil lettering (Fig. 2.16a–e)

Stencils are manufactured in clear plastic or thin metal in a wide range of lettering types. Stencilled letters are usually employed for titles, and they offer the draughtsman a simple method of producing neat uniform lettering, although it is considered by many to be mechanical and uninteresting in appearance. Possibly for this reason guided pen lettering is often preferred.

Guided pen lettering

Synonymous with the term guided pen lettering is the lettering system known as the UNO stencil (Fig. 2.17), which, because of its clarity and speed of execution, has proved itself to be a popular and extensively used medium. The special pens and plastic guides used with the system

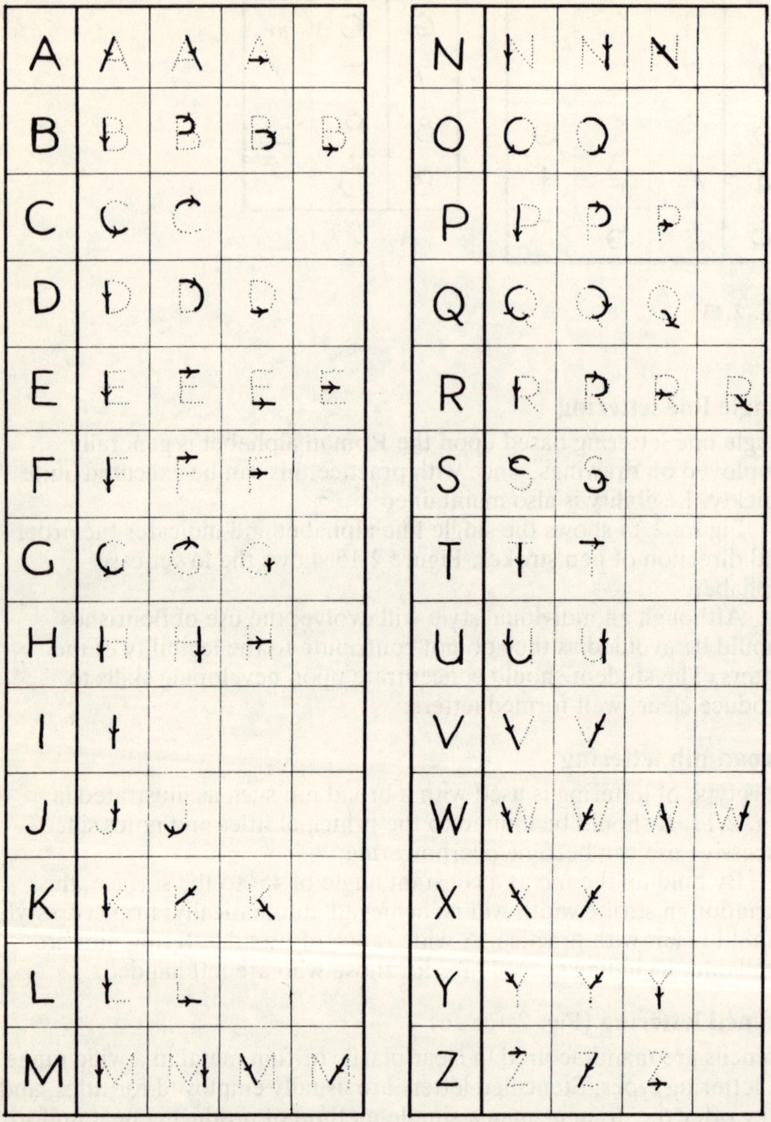

Fig. 2.14 Upper case lettering

provide clearly legible lettering and the wide range of sizes and lettering styles including upper and lower case, sloping and I.S.O. characters, making the system acceptable on all types of drawings.

The popularity of this form of lettering has been furthered by the more recent introduction of lettering guides used in conjunction with technical pens. The use of such equipment obviates the necessity to

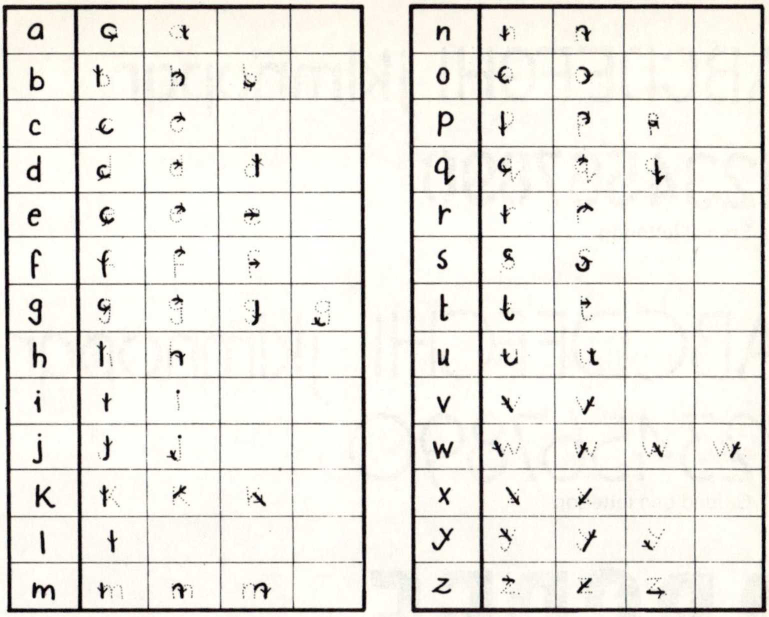

Fig. 2.15 Lower case lettering

change instruments during the course of drawing, which in turn represents a time saving.

Despite the apparent simplicity of the system a certain amount of practice is necessary to achieve an acceptable standard of lettering.

Spacing of characters is of particular importance since poorly spaced letters will result in an untidy and unsatisfactory overall presentation.

Pressure transfer lettering

The manufacture and application of pressure transfer lettering is described in Chapter 1.

When properly applied, this form of lettering is unsurpassed for its impact and sophistication, and for this reason its use is generally limited to presentation and graphic work.

To obtain the best results from the medium, particularly attention must be paid to the spacing of the letters; guide lines may be drawn to assist in the correct spacing.

Figure 2.18 represents a selection of lettering styles which are suitable for architectural work, but with the very extensive range of lettering styles and sizes which are available, the manufacturer's catalogue should be consulted to gain a full appreciation of the product range.

The application of pressure transfer lettering is shown in Fig. 2.19.

ABCDEFGHI jklmnopqr

1234567890

(a) Stencil lettering

ABCDEFGHI jklmnopqr

1234567890

(b) Guided pen lettering

ABCDEF

12345

(c) Pressure transfer lettering

ABCDEFGHIJKLM nopqrstuvwxyz

1234567890

(d) Freehand fine nib lettering

ABCDEFGHIJKLM nopqrstuvwxyz

1234567890

(e) Freehand broad nib lettering

Fig. 2.16 Types of lettering

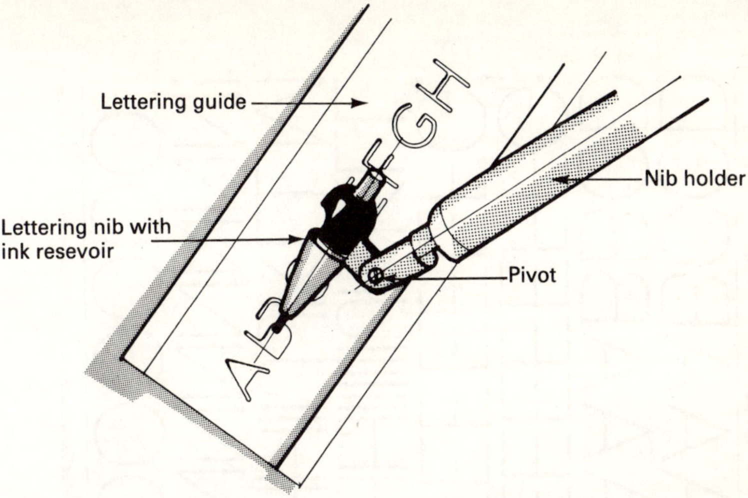

Fig. 2.17 Guided pen lettering

Notation on drawings

Explanatory notes on drawings should take the form of general notes which do not relate specifically to any part of the drawing, or, alternatively, detailed notes which refer to particular areas of the drawing.

General notes

Such notes can usually be grouped within a notes column which should be so positioned to maintain the balance of the drawing. The notes should be broken into distinct paragraphs and carefully spaced to ensure clarity and ease of reading.

Detailed notes

Notes which refer to a specific part of the drawing should be positioned near to the relevant detail, and linked to the drawing by a leader arrow. Care should be taken to avoid long or over-powering leader lines and arrow heads.

Notes are generally 2 to 4 mm high and should not be underlined unless of particular importance. Any abbreviations used in notation should accord with the relevant British Standards, and a selection of commonly encountered abbreviations is given in Table 2.1.

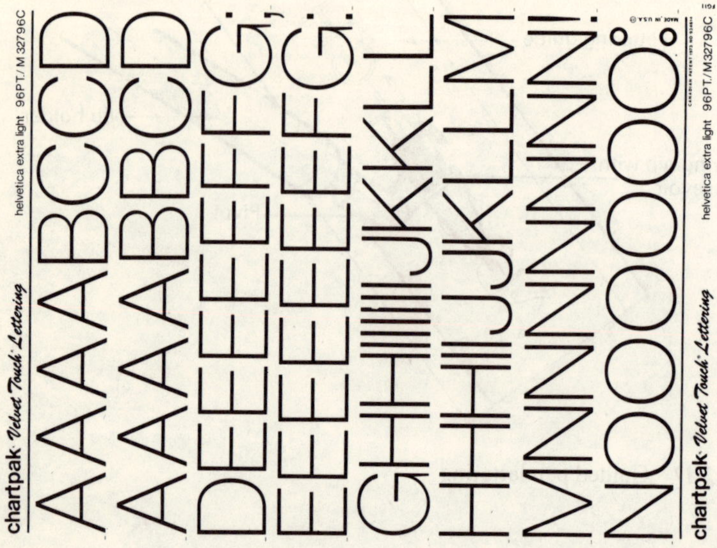
chartpak· Velvet Touch· Lettering
helvetica extra light 96PT./M 32796C

chartpak· Velvet Touch· Lettering
grotesque no.9 72PT./M 32672C

Fig. 2.18 Pressure transfer lettering

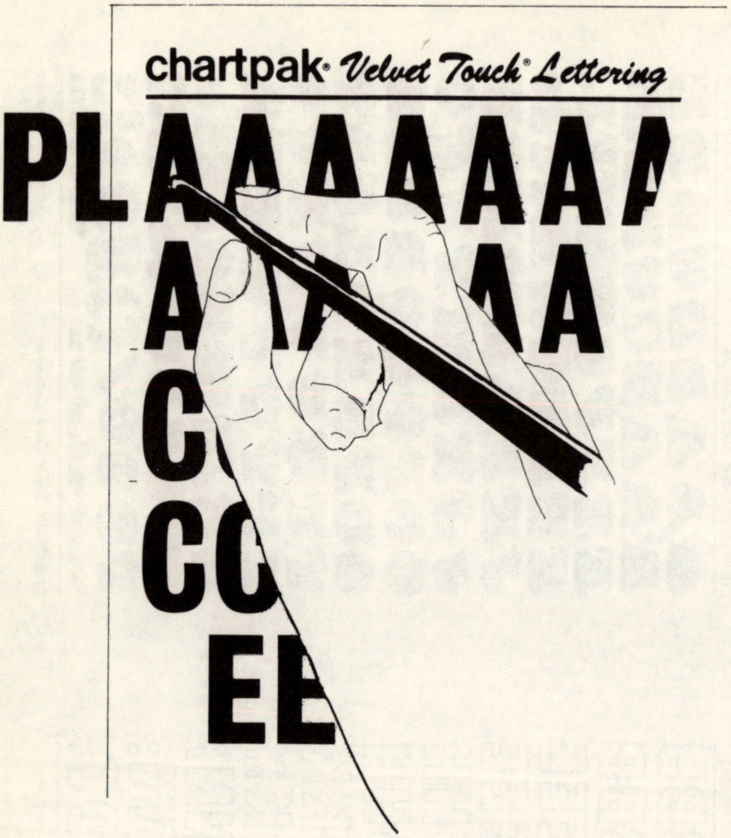

Fig. 2.19 Application of dry pressure transfer lettering

Table 2.1 Abbreviations

Aggregate	Agg.	Mild Steel	M.S.
Air Brick	A/B	Minimum	Min.
Approximate	approx.	Nominal	Nom.
Asbestos Cement	A/C	Not to Scale	N.T.S.
Asbestos Insulation Board	A.I.B.	Number	No.
Back Inlet Gully	B.I.G.	Ordnance Datum	O.D.
Bench Mark	B.M.	Ordinary Portland Cement	O.P.C.
Brickwork	Bwk.	Overall	O/all
British Standard	B.S.	Petrol Interceptor	P.I.
Cast Iron	C.I.	Radius	Rad.
Centre Line	C	Rainwater Pipe	R.W.P.
Centre to Centre	C/C	Reinforced Concrete	R.C.
Cleaning Eye	C.E.	Rodding Eye	R/E
Code of Practice	C.P.	Rolled Steel Joist	R.S.J.
Column	Col.	Rolled Steel Channel	R.S.C.
Concrete	Conc.	Self Closing	S.C.
Cover Level	C.L.	Setting-out Point	S.O.P.
Damp Proof Course	D.P.C.	Softwood	S/W
Damp Proof Membrane	D.P.M.	Soil and vent Pipe	S.V.P.
Diameter	DIA or Ø	Square	Sq.
Drawing	Dwg.	Stancheon	Stan.
Existing	Exg.	Structural Floor Level	S.F.L.
Expanded Metal lath	E.M.L.	Surface Water Sewer	S.W.S.
Extracted from	Ex.	Temporary Bench Mark	T.B.M.
Figure	Fig.	Tounged and Grooved	T & G
Finished Floor Level	F.F.L.	Trapped Gully	T.G.
Fire Resisting	F.R.	Universal Beam	U.B.
Foul Sewer	F.S.	Universal Column	U.C.
Fresh Air Inlet	F.A.I.	Vent Pipe	V.P.
Galvanized	Galv.	Volume	Vol.
Gradient	Grad.	Vertical Damp Proof Course	V.D.P.C.
Ground Level	G.L.	Wash Hand Basin	W.H.B.
Gulley	G.	Waste Pipe	W.P.
High Alumina Cement	H.A.C.	Waste and Vent Pipe	W.V.P.
High Yield Steel	H.Y.S.	Water Closet	W.C.
Inspection Chamber	I.C.	Wrought Iron	W.I.
Invert Level	I.L.	Yard Gully.	Y.G.
Lavatory Basin	L.B.		
Maximum	Max.		
Manhole	M.H.		

Chapter 3

Plane geometry

Lines

The division of lines

There are several ways in which straight lines may be geometrically divided into equal parts, and the draughtsman should be familiar with the various methods.

To bisect a line

When a line is bisected it is divided into two equal parts. The line bisecting the given line is known as the perpendicular bisector and it crosses the given line at right angles.

To bisect a line using compasses (Fig. 3.1a)
Method

1. Draw the given line AB.
2. Set compasses to any distance greater than half of AB.
3. Strike arcs from A and B to give intersections at C and D. Join C to D to give the perpendicular bisector, bisecting AB at E, such that AE = EB.

To bisect a line using a set square (Fig. 3.1b)
Method

1. Place the set square at A and B, constructing lines from both to intersect at C.
2. From C drop a perpendicular passing through the given line at E.

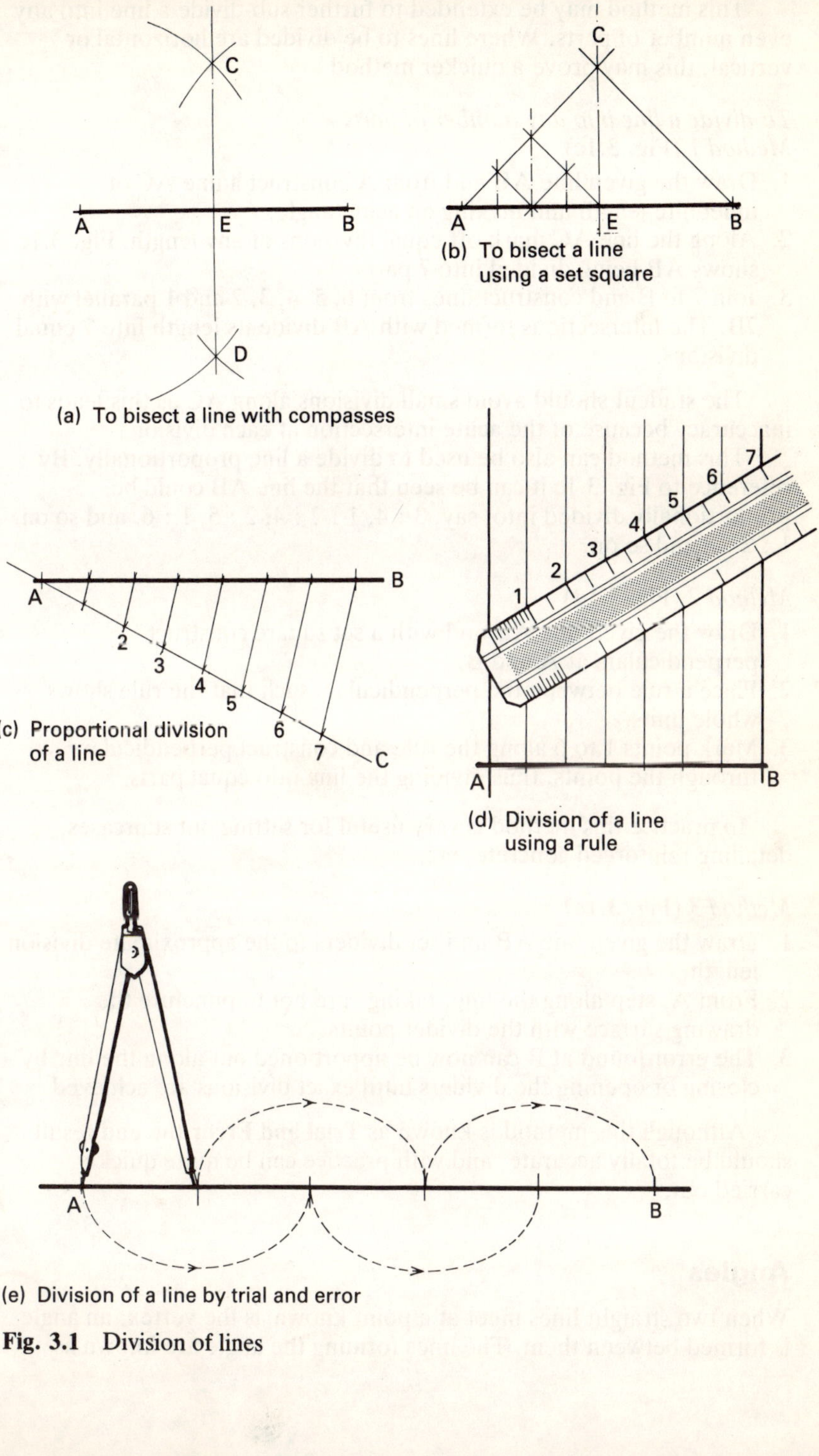

Fig. 3.1 Division of lines

This method may be extended to further sub-divide a line into any even number of parts. Where lines to be divided are horizontal or vertical, this may prove a quicker method.

To divide a line into any number of parts
Method 1 (Fig. 3.1c)

1. Draw the given line AB and from A construct a line AC of indefinite length and making an acute angle.
2. Along the line AC mark off equal divisions of any length. Fig. 3.1c shows AB being divided into 7 parts.
3. Join 7 to B and construct lines from 6, 5, 4, 3, 2 and 1 parallel with 7B. The intersections formed with AB divide its length into 7 equal divisions.

The student should avoid small divisions along AC as this leads to inaccuracy because of the acute intersection at each division.

This method can also be used to divide a line proportionally. By reference to Fig. 3.1c it can be seen that the line AB could be proportionally divided into, say, 3 : 4, 1 : 2 : 4, 2 : 5, 1 : 6, and so on. 1 : 2 : 3 and so on.

Method 2 (Fig. 3.1d)

1. Draw the given line AB and with a set square construct perpendiculars at A and B.
2. Place a rule between the perpendiculars such that the rule shows whole units.
3. Mark points 1 to 6 along the rule and construct perpendiculars through the points, thus dividing the line into equal parts.

In practice this method is very useful for setting out staircases, detailing reinforced concrete, etc.

Method 3 (Fig. 3.1e)

1. Draw the given line AB and set dividers to the approximate division length.
2. From A, step along the line, taking care not to puncture the drawing surface with the divider points.
3. The error found at B can now be apportioned out along the line by closing or opening the dividers until exact divisions are achieved.

Although this method is known as **Trial and Error** the end result should be totally accurate, and with practice can be quite quickly carried out.

Angles

When two straight lines meet at a point known as the vertex, an angle is formed between them. The lines forming the angle are known as

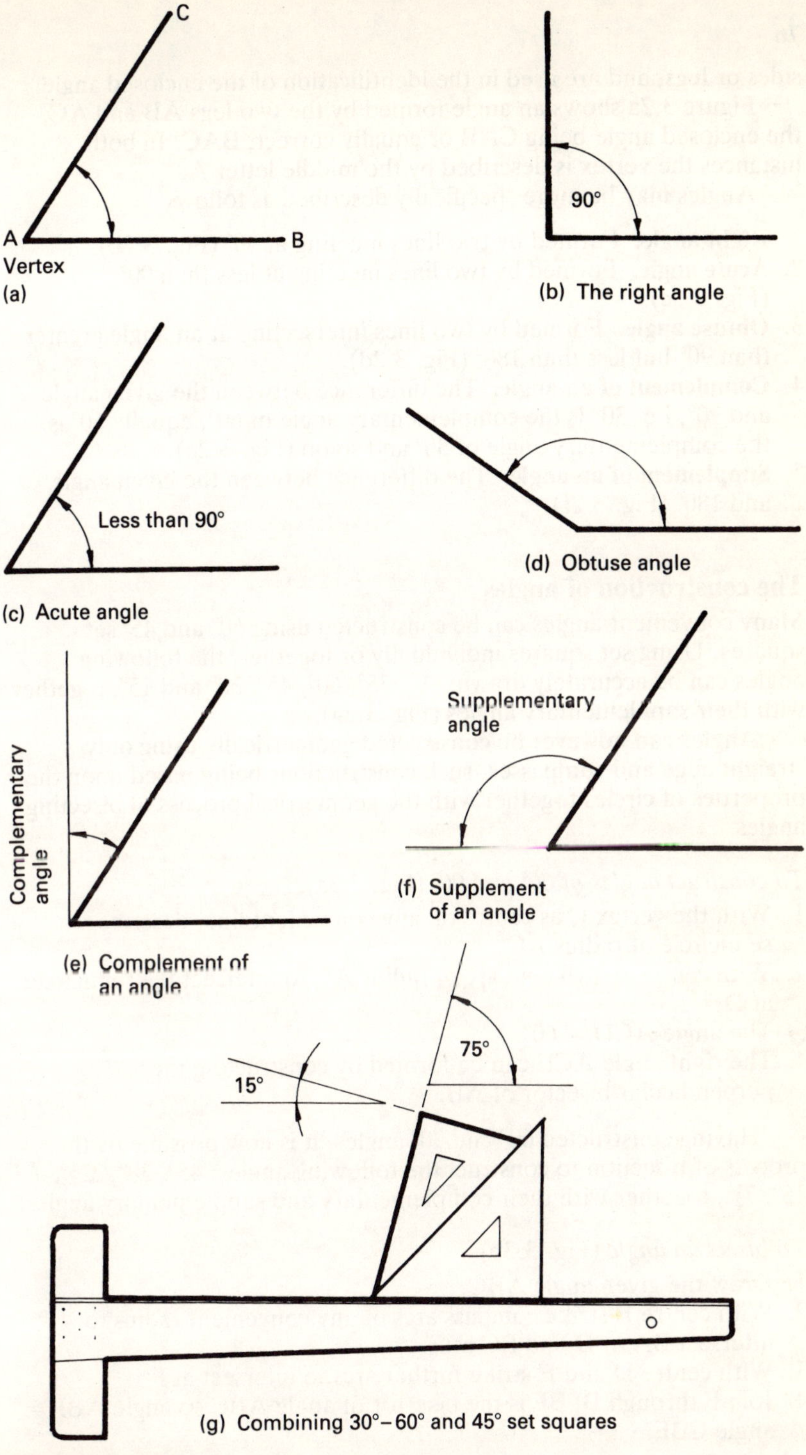

Fig. 3.2 Angles

sides or legs, and are used in the identification of the enclosed angle.

Figure 3.2a shows an angle formed by the two legs AB and AC, the enclosed angle being CAB or equally correct, BAC. In both instances the vertex is described by the middle letter A.

Angles may be more specifically described as follows:

1. **Right angle.** Formed by two lines meeting at 90° (Fig. 3.2b).
2. **Acute angle.** Formed by two lines meeting at less than 90° (Fig. 3.2c).
3. **Obtuse angle**. Formed by two lines intersecting at an angle greater than 90° but less than 180° (Fig. 3.2d).
4. **Complement of an angle.** The difference between the given angle and 90°, i.e. 30° is the complementary angle of 60°, equally 60° is the complementary angle of 30° and so on (Fig. 3.2e).
5. **Supplement of an angle.** The difference between the given angle and 180° (Fig. 3.2f).

The construction of angles

Many convenient angles can be constructed using 60° and 45° set squares. Using set squares individually or together, the following angles can be accurately drawn: 90°, 75°, 60°, 45°, 30° and 15°, together with their supplementary angles (Fig. 3.2g).

Angles can however be constructed geometrically using only straight edge and compasses, such constructions being based upon the properties of circles together with the geometrical process of bisecting angles.

To construct angles of 60° and 90° (Fig. 3.3a).

1. With the vertex C as centre on any convenient line, describe a semicircle of radius AC.
2. With centre A strike an arc of radius AC to intersect the semicircle at D.
3. The angle ACD = 60°.
4. The right angle ACE can be formed by constructing the perpendicular bisector of AB.

Having constructed 60° and 90° angles, it is now possible by the process of bisection to construct the following angles: 45°, 30°, $22\frac{1}{2}$°, 15°, $7\frac{1}{2}$°, together with their complementary and supplementary angles.

To bisect an angle (Fig. 3.3b)

1. Draw the given angle ABC.
2. With centre B strike compass arcs of any convenient radius to intersect BA at D and BC at E.
3. With centre D and E strike further arcs to intersect at F.
4. Join F through B; BF is the bisector of angle ABC so angle ABF = angle CBF.

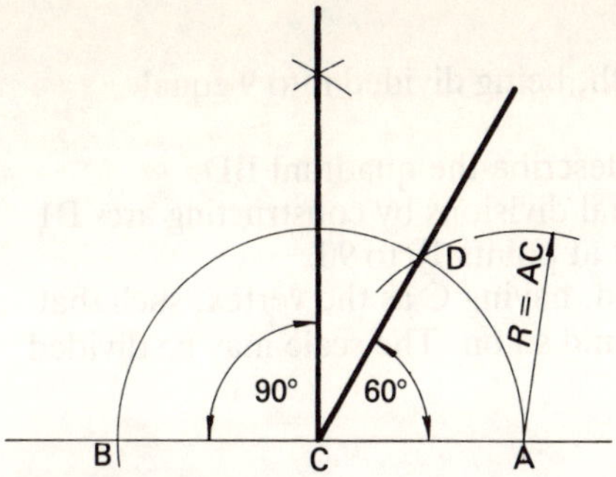

a) To construct angles of 60° and 90°

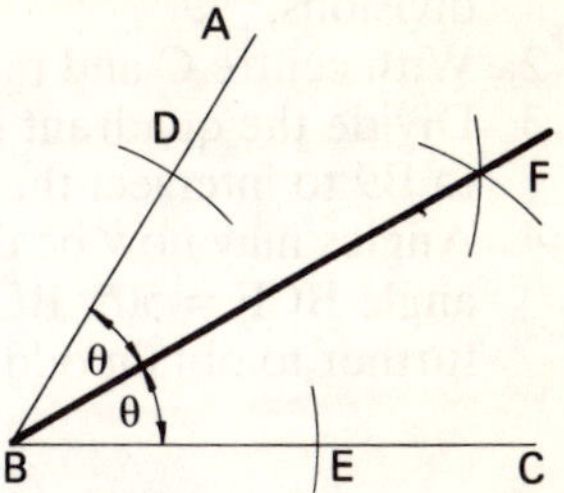

(b) To bisect a given angle

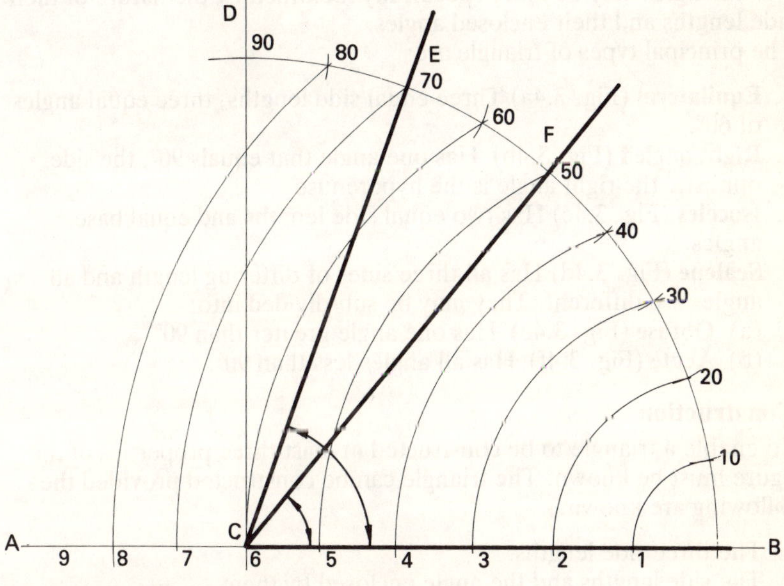

(c) Scale of chords

Fig. 3.3

To gain the maximum degree of accuracy the compass setting should not be of too small a radius.

The scale of chords (Fig. 3.3c)

The scale of chords provides a method of measuring and constructing angles using only a rule and compass.

Method

1. Draw line AB of suitable length, being divided into 9 equal divisions.
2. With centre C and radius CB describe the quadrant BD.
3. Divide the quadrant into 9 equal divisions by constructing arcs B1 to B9 to intersect the quadrant at points 10 to 90.
4. Angles may now be determined, having C as the vertex, such that angle BCF = 50°, BCE = 70° and so on. The scale may be divided further to obtain odd degrees.

Triangles and circles

A triangle may be defined as a plane figure having three sides and enclosing three angles, the sum of these angles being 180°.

Triangles may be more specifically identified by the nature of their side lengths and their enclosed angles.
The principal types of triangle are:

1. **Equilateral** (Fig. 3.4a) Three equal side lengths, three equal angles of 60°.
2. **Right angled** (Fig. 3.4b) Has one angle that equals 90°; the side opposite the right angle is the hypotenuse.
3. **Isoceles** (Fig. 3.4c) Has two equal side lengths and equal base angles.
4. **Scalene** (Fig. 3.4d) Has all three sides of differing length and all angles are different. They may be sub-divided into:
   (a) **Obtuse** (Fig. 3.4e) Has one angle greater than 90°.
   (b) **Acute** (Fig. 3.4f) Has all angles less than 90°.

Construction

To enable a triangle to be constructed at least three properties of the figure must be known. The triangle can be constructed provided the following are known.

1. The three side lengths.
2. The side lengths and the angle enclosed by them.
3. One side length and two of the angles.

Now consider the methods of construction for each case mentioned above:

1. Construct a triangle having sides of 50 mm, 40 mm and 70 mm (Fig. 3.5a).
   1. Select any one side and draw the base of the triangle to the given length AB.
   2. Set compasses to either of the remaining side lengths and describe an arc about A.

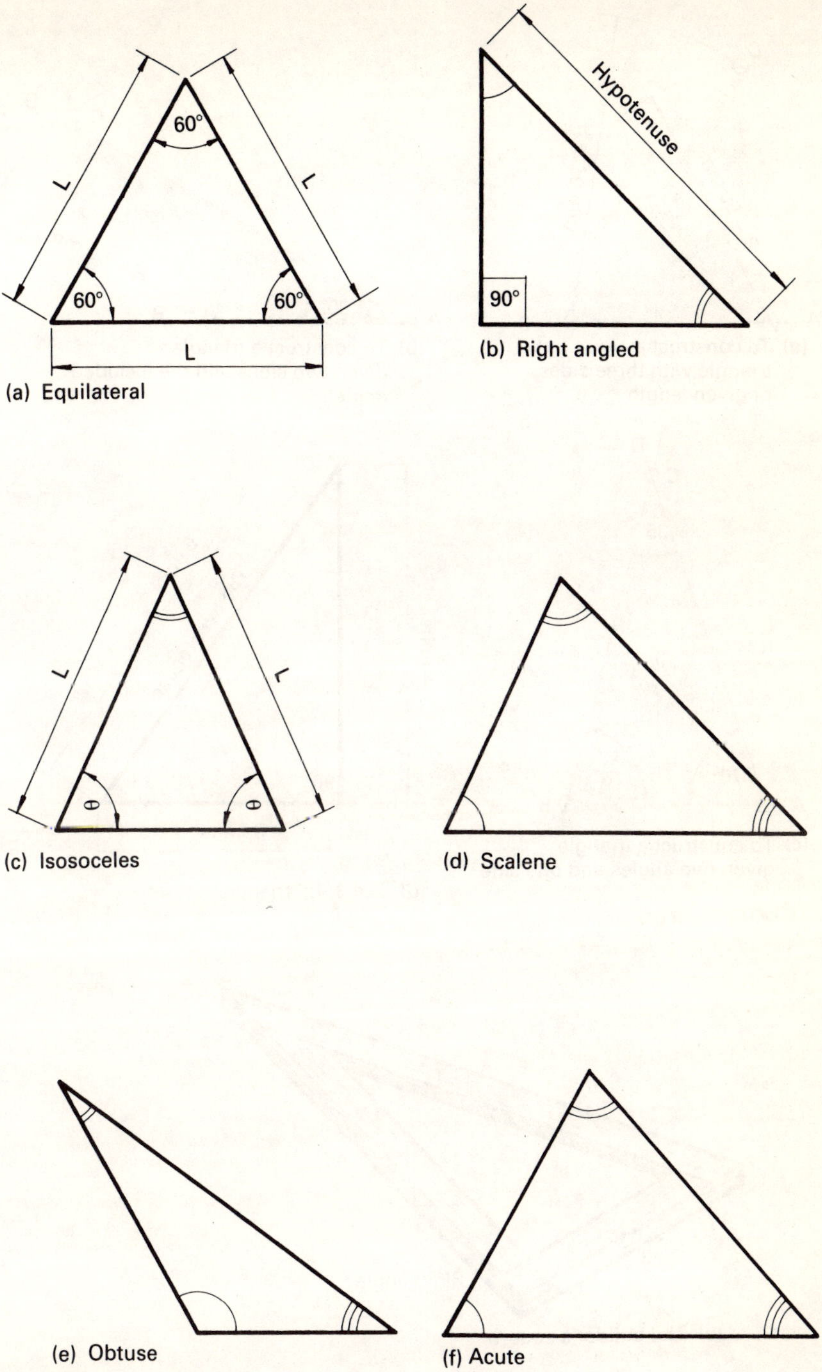
60°
60°
60°
L
L
L
(a) Equilateral
Hypotenuse
90°
(b) Right angled
L
L
θ
θ
(c) Isosoceles
(d) Scalene
(e) Obtuse
(f) Acute

Fig. 3.4

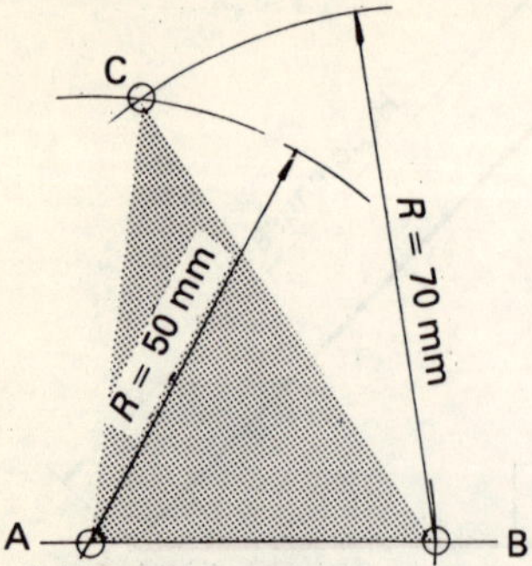

(a) To construct a triangle with three sides of given length

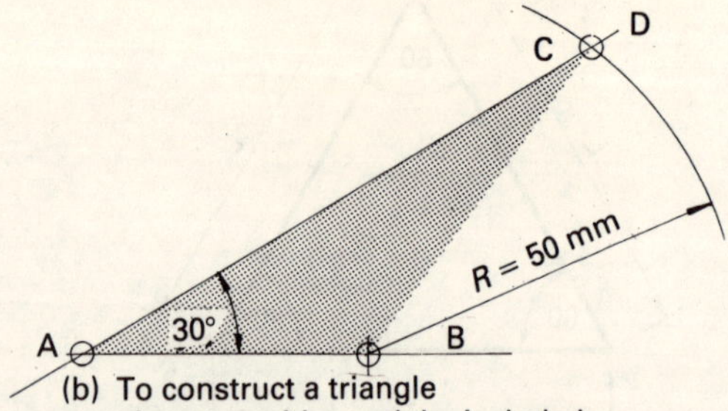

(b) To construct a triangle given two sides and the included angle

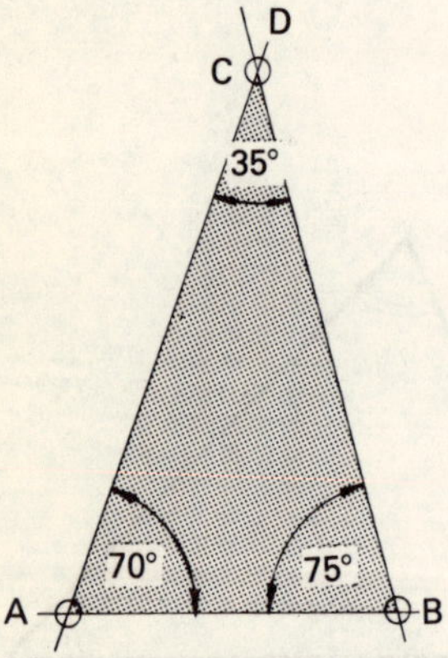

(c) To construct a triangle given two angles and one side

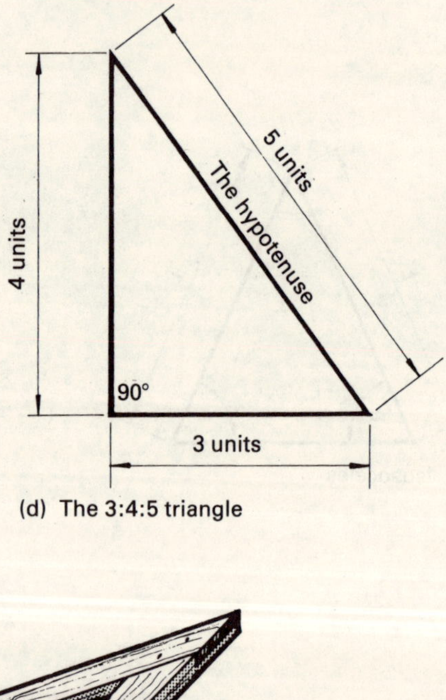

(d) The 3:4:5 triangle

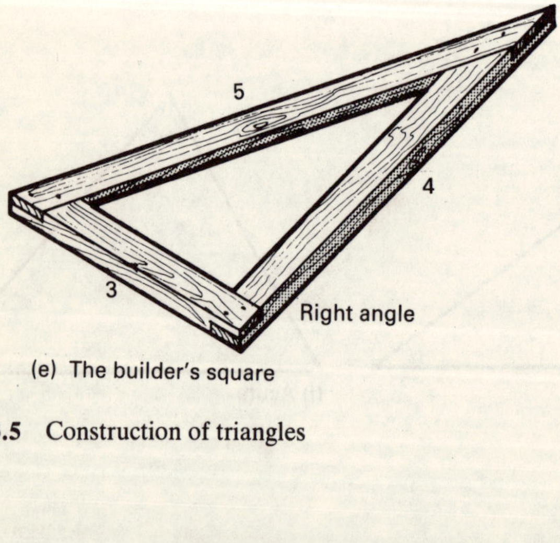

(e) The builder's square

Fig. 3.5 Construction of triangles

3. Set compasses to remaining side length and describe an arc about B to intersect previous arc at C.
4. Join AB and C to produce the required triangle.

2. Construct a triangle of sides 40 mm and 50 mm enclosing an angle of 30° (Fig. 3.5b).
   1. Draw one given side AB.
   2. From A construct a line of indefinite length to D, making an angle of 30° with AB.
   3. Set compasses to remaining side length and with centre A strike an arc intersecting AD at C.
   4. Join AC and CB to produce the required triangle.
3. Construct a triangle of side 40 mm, base angle 70° and apex angle of 35° (Fig. 3.5c).
   1. Draw the given side AB.
   2. From A construct a line of indefinite length to D, making an angle of 70° with AB.
   3. From B construct a line making angle of 75°, [180° – (70° + 35°)] with AB and intersecting AD at C.
   4. ABC is the required triangle.

The 3 : 4 : 5 triangle (Fig. 3.5d)
Of particular interest to the builder and civil engineer is the property of the 3 : 4 : 5 triangle. The significance of the figures lies in the RATIO of the side lengths in that, by pythagoras, a triangle with this side relationship must be a right-angled triangle, viz.

$$3^2 + 4^2 = 5^2 \quad \text{or} \quad 9 + 16 = 25$$

It can been seen therefore that any triangle, regardless of its actual size dimensions, will be a right-angled triangle provided that the sides retain this relationship. Therefore triangles having side lengths of 300, 400 and 500 mm, and 600, 800 and 1000 mm are all right-angled, in each case the right angle being opposite the longest side.

In practice, considerable use is made of the 3 : 4 : 5 triangle in setting out buildings using only a tape, as well as in the making of a builder's square (Fig. 3.5e).

The circle

The circle may be defined as *a plane figure bounded by a curve which maintains a constant distance from a single point called the centre.*

It is important that the student should be familiar with the terminology used in the geometry of circles (Fig. 3.6).

Circumference The curve bounding the circle, being a constant distance from the centre.
Arc A part of the circumference.

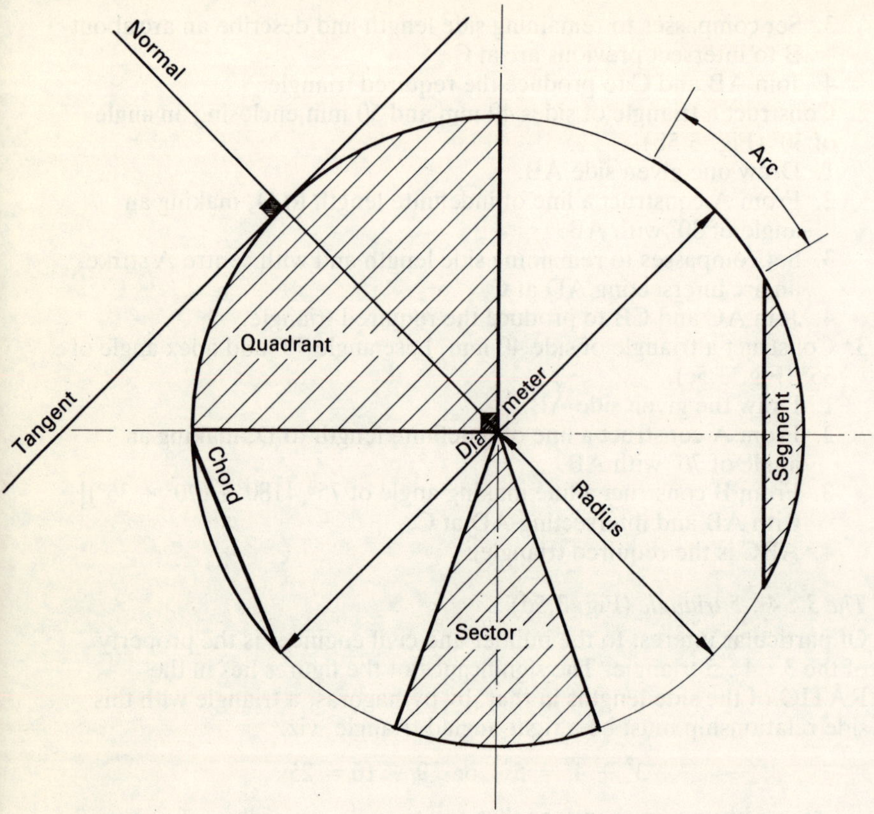

Fig. 3.6

Diameter A straight line passing through the centre of the circle, terminating at the circumference at both ends.
Radius A straight line from the centre to the circumference, i.e. half the diameter.
Normal A line drawn from the centre and extended through the circumference.
Quadrant A quarter of a circle, both in shape and area, bounded by two radii enclosing an angle of 90°.
Sector A part of a circle bounded by two radii enclosing an angle of less than 180°.
Chord A straight line, shorter than the diameter, terminating at both ends at the circumference.
Segment A portion of a circle contained by the chord and its arc.
Tangent A straight line which touches the circumference of the circle at one point, and makes an angle of 90° with the normal at that point.

To find the centre of a given circle (Fig. 3.7a)

1. Construct any two chords AB and BC.
2. Construct the perpendicular bisectors of the two chords and extend the bisector to intersect at O.
   The point O is the required centre.

To construct a circle to pass through three given points (Fig. 3.7b)

1. Set out the given points A, B and C.
2. Join A, B and C and construct the perpendicular bisectors of AB and BC.
3. Extend bisectors to intersect at O.
4. Set compasses to radius AO and describe the circle through A, B and C.

To construct a series of circular arcs to pass through a number of given points (Fig. 3.7c)

1. Set out the given points ABCD and E.
2. Join the points and construct the perpendicular bisector of each.
3. Extend the bisectors of AB and BC to intersect at O, this being the centre of the arc passing through AB and C.
4. Join O through C to intersect the bisector of CD at O1, this being the centre of the arc passing through C and D.
5. Join O1 through D to intersect the bisector of D, E at O2. Similarly, O2 is the centre of the arc passing through D and E.

To construct a circular arc of given radius to touch two intersecting lines (Fig. 3.8a)

1. Let AB and AC be lines of indefinite length intersecting at A.
2. Select at least two points along AB and two points along AC, and from these points erect perpendiculars at 1, 2, 3 and 4.
3. With compasses set to given arc radius describe arcs about 1, 2, 3 and 4 intersecting at 1_1 2_1 3_1 and 4_1.
4. Join 1_1 2_1 and 3_1 4_1 to intersect at D.
5. D is the centre of the arc of radius R which will touch AB and AC tangentially.

When curves of this nature are drawn, it is preferable to draw the final curve onto setting out lines, and then lead away from the curve with the finished straight line. The effect of this is to allow a little latitude in accuracy and avoid the too common fault of curve and straight not meeting.

In practice tangency occurs frequently when designing estate roads, detailing components, location, drawings and site plans.

To construct a circle inside a given triangle (Fig. 3.8*b*)

1. Draw the given triangle ABC.
2. Construct the bisectors of two of its angles.

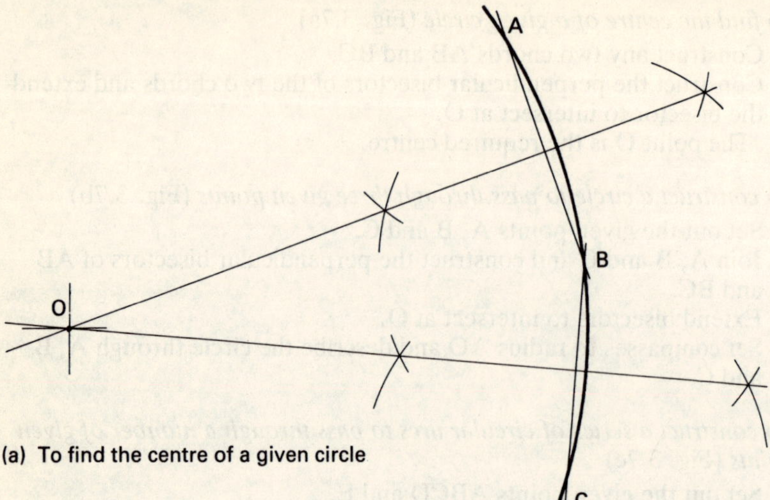

(a) To find the centre of a given circle

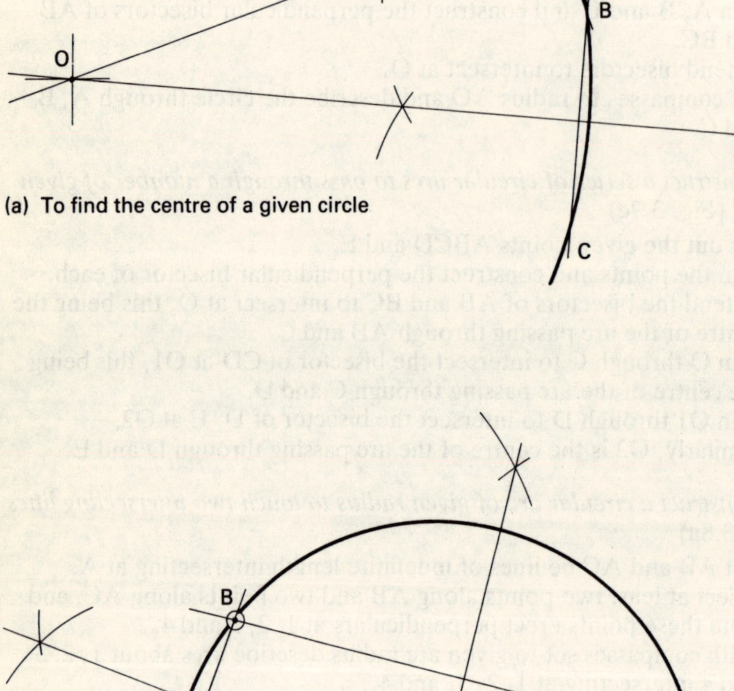

(b) To construct a circle to pass through three given points

Fig. 3.7

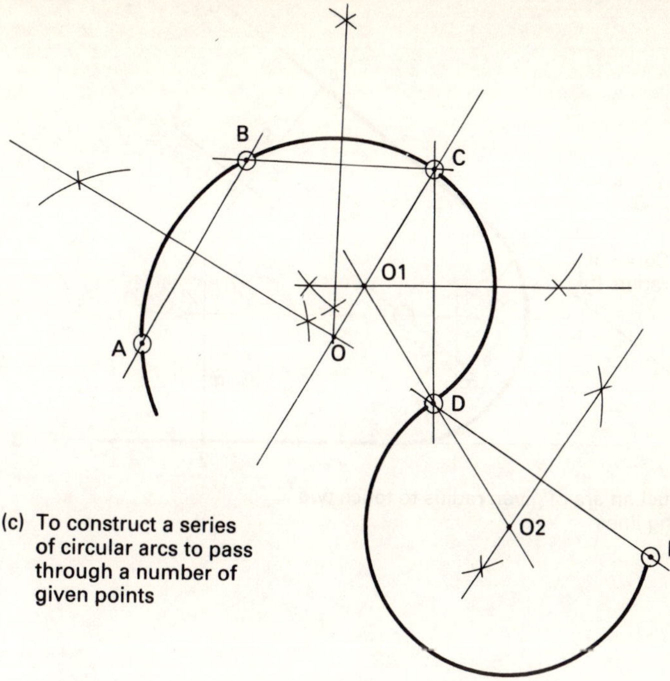

(c) To construct a series of circular arcs to pass through a number of given points

3. Extend the bisectors to intersect at O.
4. From O construct a perpendicular to one side of the triangle OD.
5. With radius OD describe the inscribed circle to touch all sides of the triangle.

To construct a circle about a given triangle (Fig. 3.8*c*)

1. Draw the given triangle ABC.
2. Construct the perpendicular bisector of two of the sides of the triangle to intersect at O.
3. With radius AO describe the circumscribed circle to pass through points AB and C.

   It can be seen that this construction is similar to that in Fig. 3.7c.

Tangents

A tangent, as previously described, is a straight line which touches the circumference of a circle at a point, making an angle of 90° with the normal at that point.

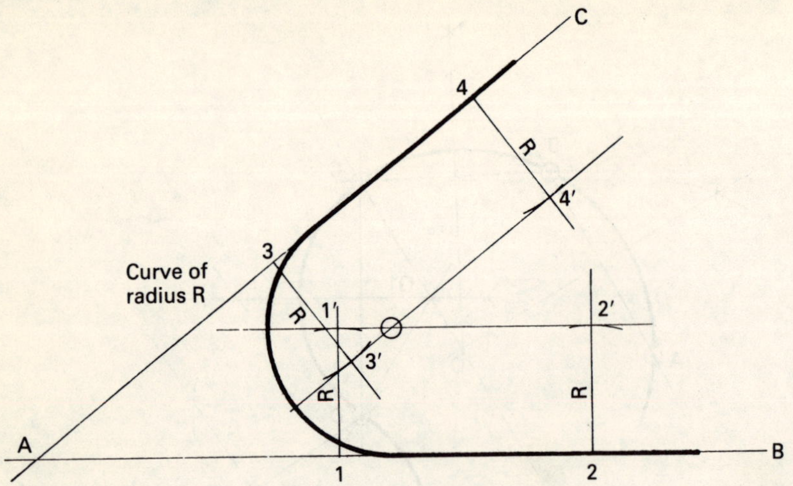

(a) To construct an arc of given radius to touch two intersecting lines

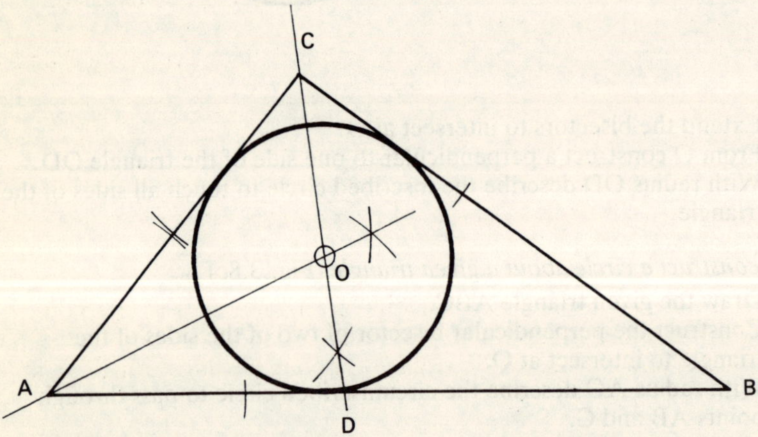

(b) To construct the inscribed circle

Fig. 3.8

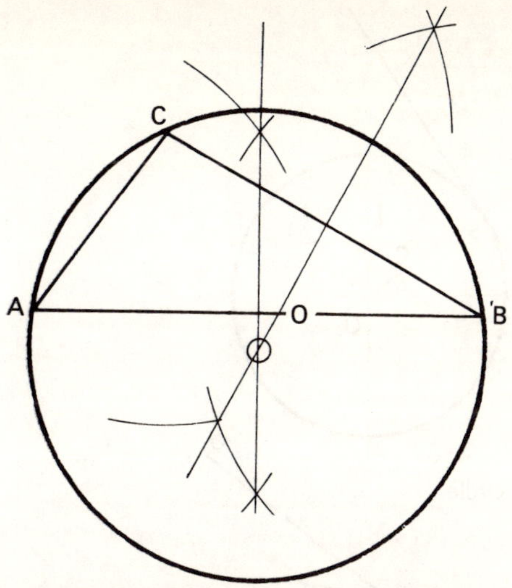

(c) To construct the circumscribed circle

To construct the tangent to a point on a circle (Fig. 3.9a)

1. Select point P on the circumference of the given circle.
2. Construct a line OA of indefinite length from centre O passing through the point P.
3. With centre P construct arcs of any radius to intersect OA at B and C.
4. Construct the perpendicular bisector of BC which is the required tangent.

To construct the tangent to a point on an arc of unknown centre (Fig. 3.9b)

1. Let P be the point of contact.
2. With P as centre describe an arc of convenient radius to intersect the given arc at A and B.
3. Join AB and construct CD parallel to it, and passing through point P.

To construct a tangent to a circle from a point outside the given circle (Fig. 3.9c)

1. Let O be the centre of the given circle and A the given point.
2. Join OA and construct its perpendicular bisector at B.
3. With centre B and radius AB describe an arc to intersect the circumference of the given circle at P.
4. Join AP, which is the required tangent.

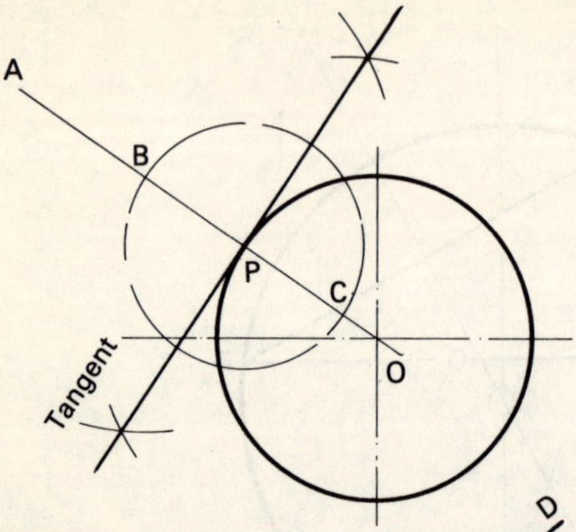

(a) Tangent to a circle

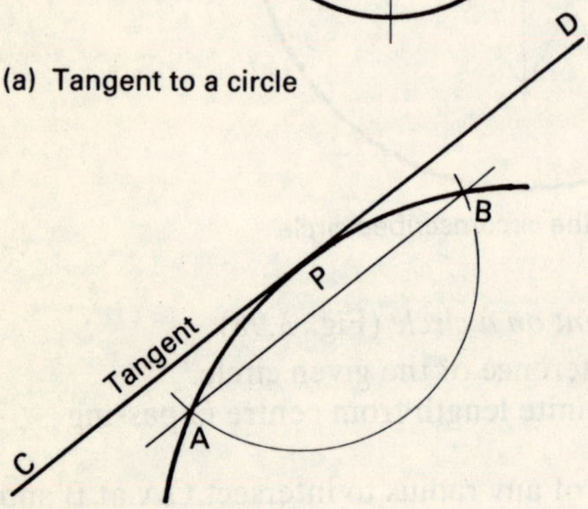

(b) Tangent to an arc

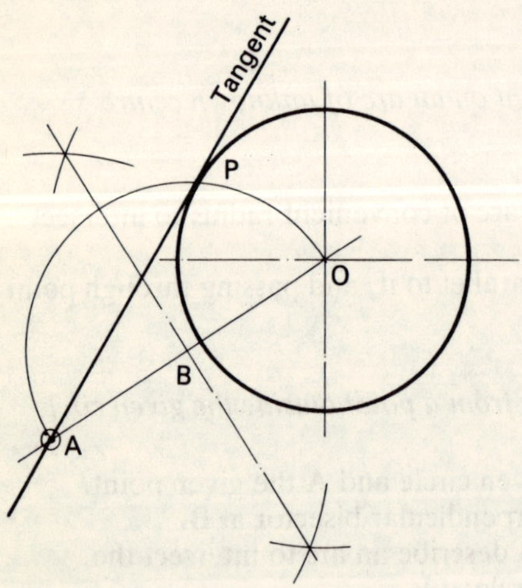

(c) Tangent to a circle from a given point

Fig. 3.9

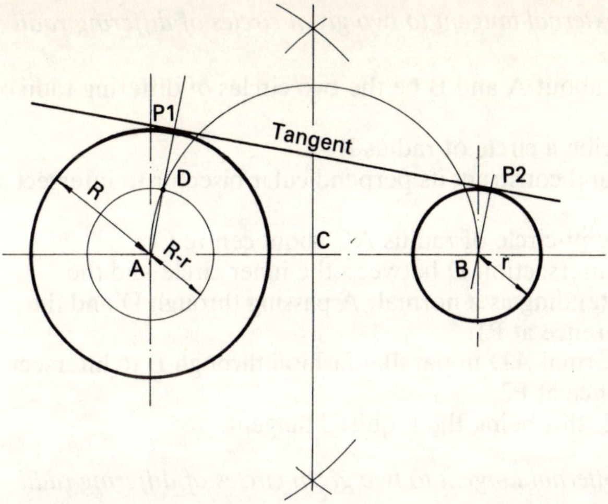

(d) External tangent to two differing circles

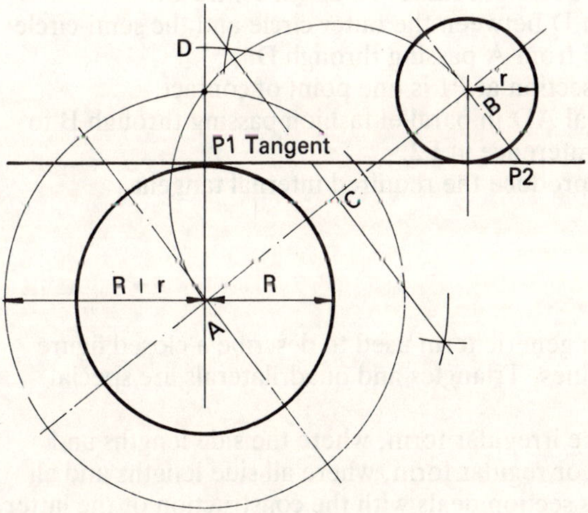

(e) Internal tangent to two differing circles

To construct the external tangent to two given circles of differing radii (Fig. 3.9d)

1. Let the circles about A and B be the two circles of differing radii of R and r.
2. About A describe a circle of radius R−r.
3. Join A and B and construct its perpendicular bisector to intersect at C.
4. Describe the semi-circle of radius AC about centre C.
5. Join A to the intersection D between the inner circle and the semi-circle, extending as a normal, A passing through D and the outer circumference at P1.
6. Transfer the normal AD in parallel fashion through B to intersect the circumference at P2.
7. Join P1 and P2, this being the required tangent.

To construct the internal tangent to two given circles of differing radii (Fig. 3.9e)

1. Let the circles about A and B be the two circles of differing radii of R and r.
2. About A describe a circle of radius R + r.
3. Join A and B and construct its perpendicular bisector to intersect at C.
4. Describe the semi-circle of radius AC about centre C.
5. At the intersection D between the outer circle and the semi-circle construct a normal from A passing through D.
6. The point of intersection at P1 is one point of contact.
7. Transfer the normal AD in parallel fashion passing through B to intersect the circumference at P2.
8. Join P1 and P2 to produce the required internal tangent.

Polygons

The term polygon is a generic term used to describe a closed figure bounded by straight lines. Triangles and quadrilaterals are special forms of polygon.

Polygons may take irregular form, where the side lengths and angles enclosed vary, or regular form, where all side lengths and all angles are equal. This section deals with the construction of the latter.

Regular polygons are described specifically by the number of sides they possess.

Pentagon	–	five sides
Hexagon	–	six sides
Heptagon	–	seven sides
Octagon	–	eight sides
Nonagon	–	nine sides

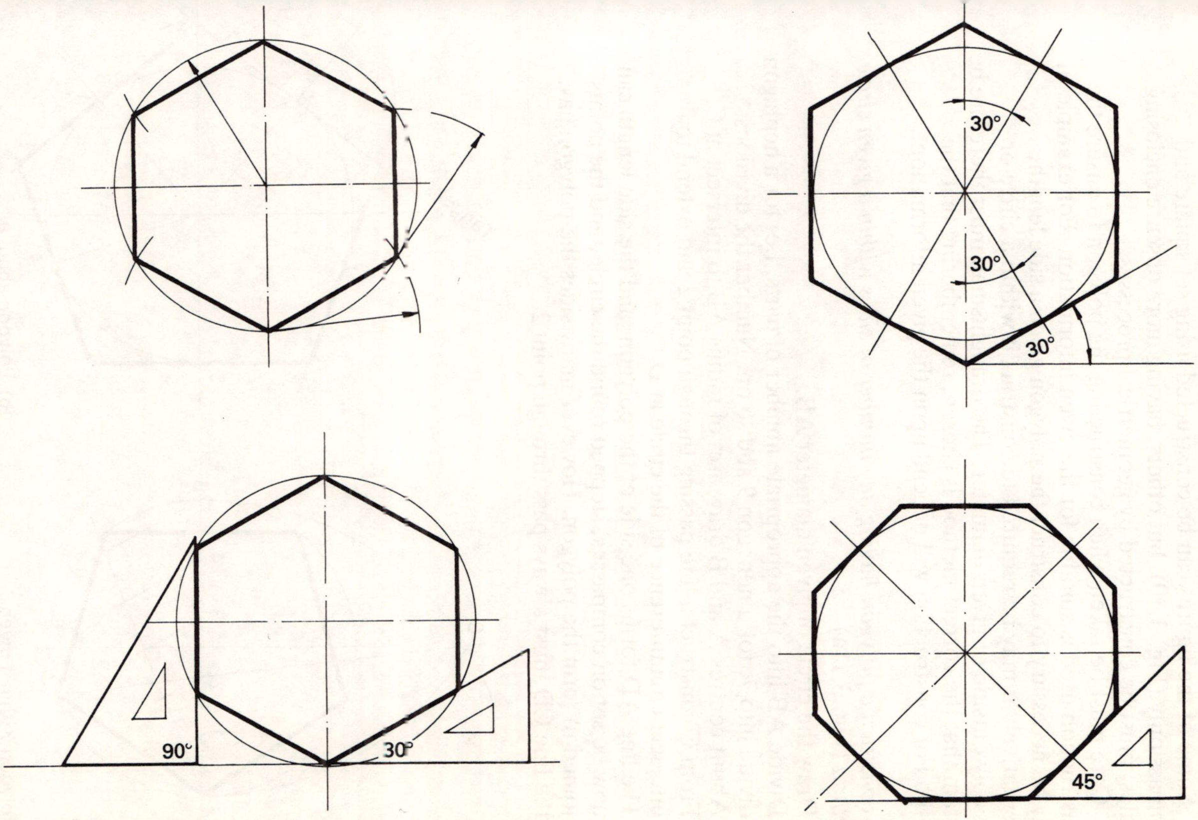

Fig. 3.10 Polygons using compass and square only

Decagon – ten sides
Undecagon – eleven sides
Dodecagon – twelve sides

Some of the figures can be constructed using set square and compass only (Fig. 3.10), but others, having more obscure enclosing angles, must be constructed by geometrical processes.

Care must be taken when constructing the polygon to ensure the construction used is suitable for the given information. For example, it may be necessary to construct the polygon given a side length; alternatively it may be required to construct it within a circle or possibly outside of given diameter. The draughtsman must therefore be certain that the correct method is chosen, remembering that the method of construction will depend upon the given information.

To construct a polygon having any number of sides within a given circle
Method (Fig. 3.11a)

1. Draw the circle of given diameter AB.
2. Divide AB into the appropriate number of parts, i.e. for a heptagon divide into 7, for a nonegon 9 and so on. Number the divisions.
3. About centres A and B draw arcs of radius AB to intersect at C.
4. From C construct a line passing through point 2 and extend to intersect circumference of the circle at D.
5. The line AD forms one side of the polygon, and the side length can now be set on compasses, stepped round the circle, and the points joined to form the polygon. However many sides the polygon has, the line CD must always pass through point 2.

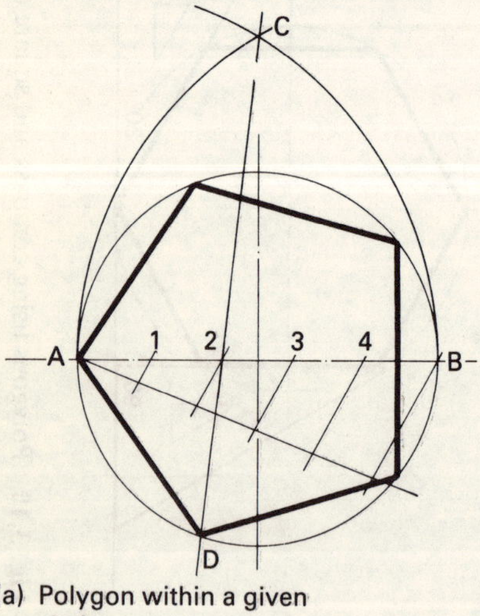

(a) Polygon within a given circle

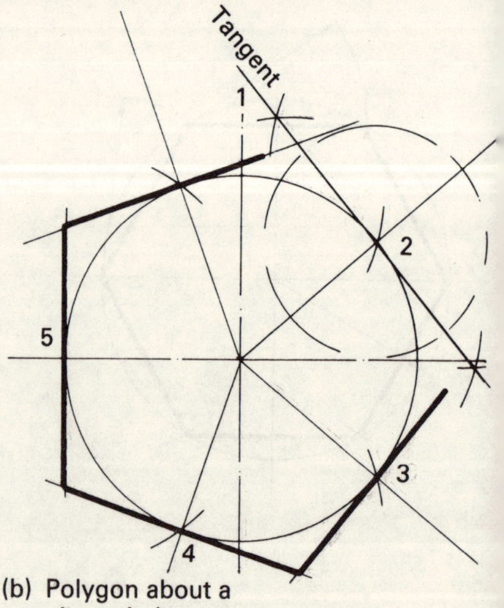

(b) Polygon about a given circle

Fig. 3.11

To construct a polygon about a given circle
Method (Fig. 3.11b)

1. Draw the given circle and divide the circumference into the appropriate number of parts.
2. Draw normal through the points and construct the tangent to the normal. It is often unnecessary to construct all the tangents where certain sides of the polygon are parallel and lines can be transferred.

To construct a regular polygon having any number of sides on a given side length
Method 1 (Fig. 3.12a)

1. Draw the given side AB and construct its perpendicular bisector.
2. From A construct a line 45° to intersect the centre line at 4. The point 4 is the centre of a circle which will circumscribe a 4-sided figure (a square) of side AB as A4 is equal to half the diagonal length of the square.
3. From B construct a line at 60° to intersect the centre line at 6. The point 6 is the centre of a circle which will circumscribe a 6-sided figure (hexagon) of side AB since A6 forms half the corner-to-corner distance of the hexagon.
4. Bisect 4–6 to give point 5. This point will, for practical purposes, be the centre of the circle circumscribing a 5-sided figure (pentagon).
5. Set compasses to the distance 45 and from point 6 step up the centre line to locate points 7, 8, 9, 10, 11 and 12. Points 7 to 12 are the centre points of circles circumscribing their respective polygons, in the same way as points 4, 5 and 6.

Having constructed the circumscribing circle, the points of the polygon can be plotted by stepping around the circle, the side length AB.

Although this method is very useful to the draughtsman, it must be appreciated that it is not totally accurate, the error becoming more evident as the number of sides increases.

The reason for the error can be simply illustrated by calculation:

If AC = 20 mm
and C4 = 20 mm

$$C6 = AC \tan 60 = 34.641 \text{ mm}$$

$$C5 = C4 + \frac{4/6}{2}$$

$$= 20 + \frac{14.64}{2}$$

$$= 27.32 \text{ mm}$$

$$\therefore \text{angle } CA5 = \tan^{-1}\frac{C5}{AC}$$

$$= 53.793°$$

But the true angle so made in a pentagon should be 54°.

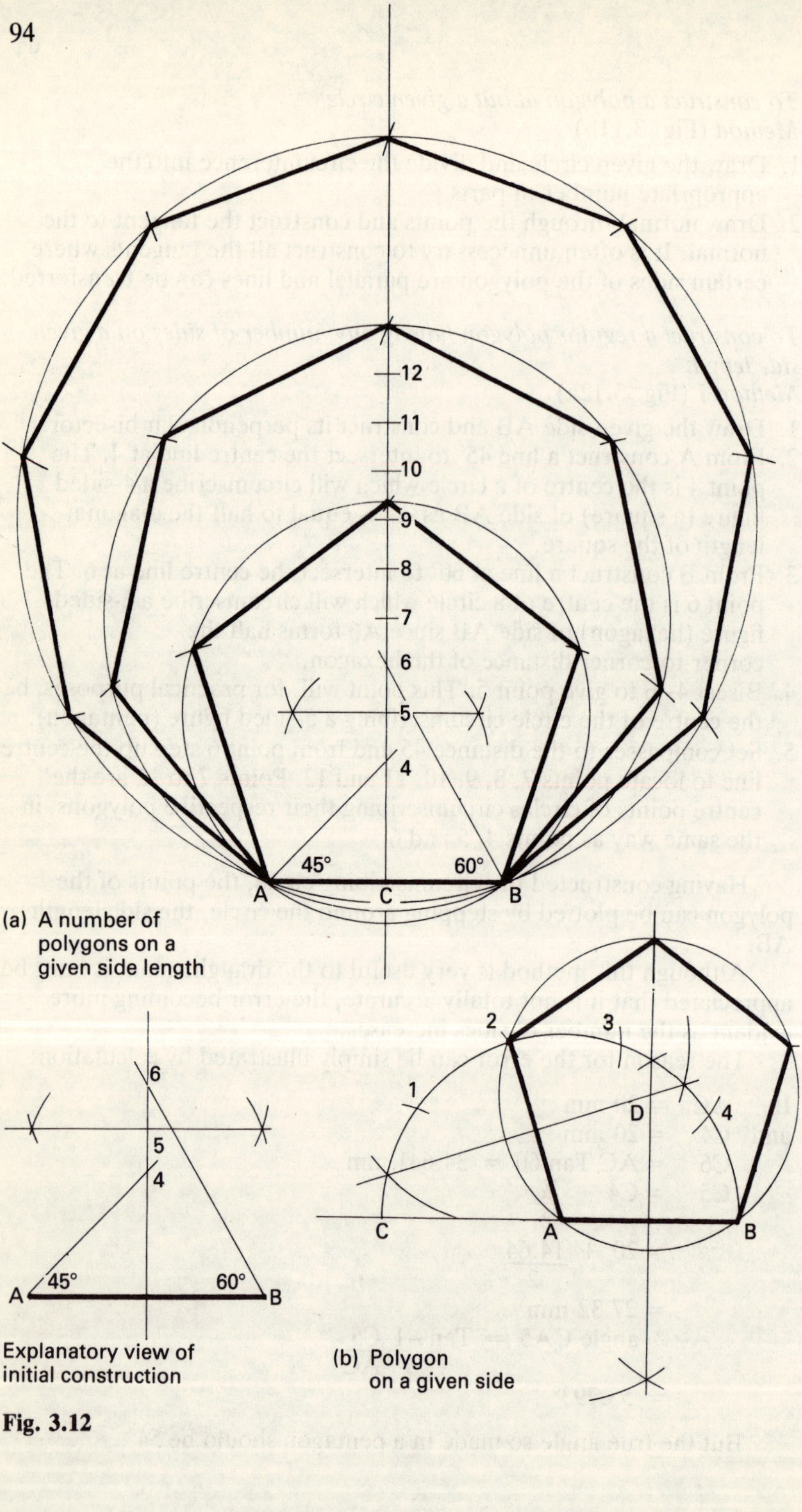

(a) A number of polygons on a given side length

Explanatory view of initial construction

(b) Polygon on a given side

Fig. 3.12

As any error becomes more noticeable, it may be accurately apportioned out using dividers.

Method 2 (Fig. 3.12b)

1. Draw the given side AB and with compasses set to radius AB and A as centre construct a semi-circle, extending AB to intersect at C.
2. Divide the semi-circle, by trial and error, into the appropriate number of divisions depending on the polygon to be drawn, i.e. 7 for heptagon, 9 for nonagon and so on. Number the divisions.
3. From A draw a line passing through point 2 on the semi-circle. A2 will be equal in length to AB, as they are both radii, and will also form a second side to the polygon.
4. Construct the perpendicular bisectors of AB and A2 to intersect at D.
5. Point D is the centre of the circle which will circumscribe the polygon. The circle can now be drawn with a radius of DA, and will pass through points A, B and 2.
6. With compasses set to AB step around the circle to establish the position of the remaining points on the polygon.

The line forming the second side of the polygon must always pass through point 2 irrespective of the number of sides the polygon has.

The ellipse

The ellipse is widely used in building and architecture, particularly in the construction of arches. It also occurs frequently in drawing practice when representing circles in various projections, especially isometric (See Ch. 6).

The ellipse can be considered in two ways, either

1. As the locus (path) of a moving point,
   or
2. As a conic section (see Ch. 5).

There are several methods of constructing true ellipses and this chapter will firstly consider the definition of an ellipse, which is: *The locus of a point which moves in such a way that the sum of the distances from two fixed points (known as focal points) is constant and is equal to the length of the major axis.*

The ellipse lies symmetrically about two axes of differing lengths, the longer being the **major axis** and the shorter the **minor axis**. The axes cross at right angles (Fig. 3.13a), each a perpendicular bisector of the other.

The **focal points** lie on the major axis and their position is established by measuring the semi-major axis length about B. This can

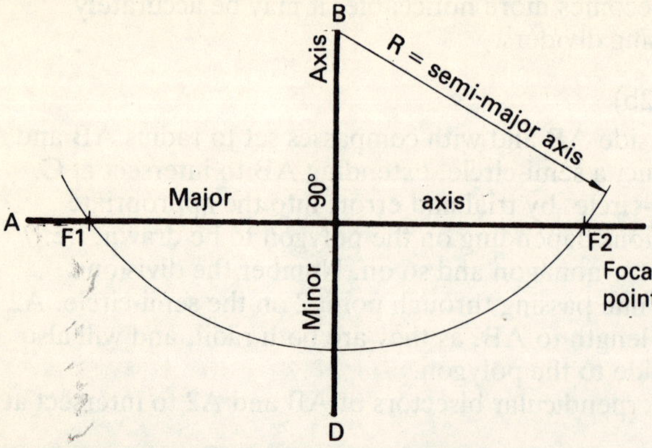

(a) To determine the focal points of an ellipse

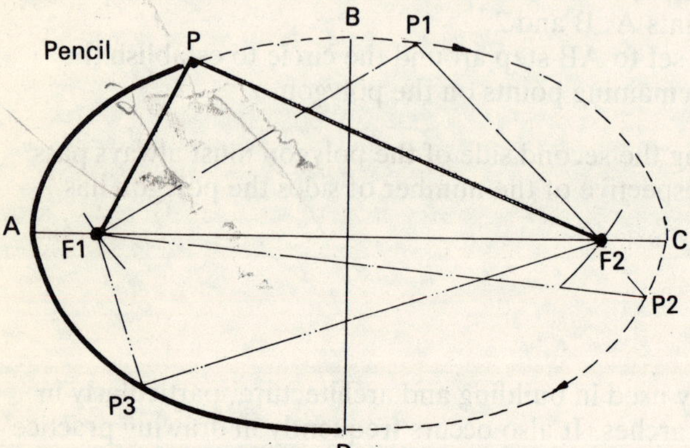

(b) To construct an ellipse by pins and string

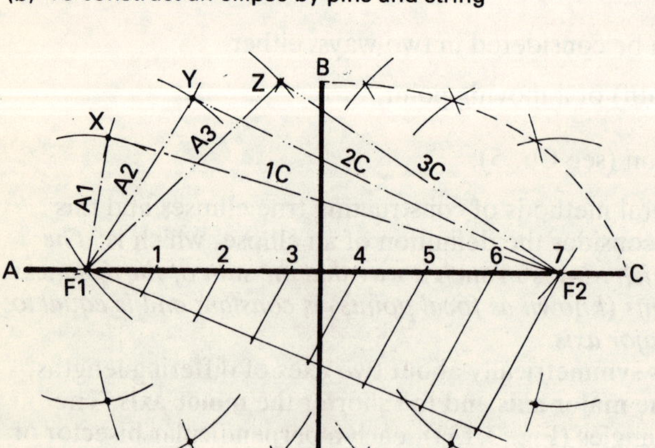

(c) To construct an ellipse by intersecting arcs

Fig. 3.13

be done simply by setting compasses to OC and, with B as centre, draw an arc to intersect the major axis of F1 and F2.

By reference to the definition, F1B + F2B = AC as B is a point on the ellipse.

In the same way any point on the ellipse will satisfy this equation.

Select any point P which lies on the curve, then F1P + F2P = AC. This is true for any point which lies on the elliptical curve.

This basic property of the ellipse can be simply proven using paper pins and string.

To construct an ellipse using pins and string

Method (Fig. 3.13b)

Draw major and minor axes on the paper and mark focal points. As described earlier, take a piece of string equal in length to the major axis and fix its ends to F1 and F2. By pulling the string tight with pencil P and tracking the pencil over the paper, keeping the string tight, the path of the ellipse will be traced out. The string length cannot alter so any point on the trace must satisfy the definition in that the sum of the distances F1P + F2P must equal the length of the string which in turn equals AC. Again this is true for any point P1, P2, etc. on the curve. The method described does not serve the draughtsman adequately, although it may be used for setting out and it does illustrate the basic property of the ellipse.

There are several ways in which the ellipse can be geometrically constructed, the principal methods being:

(a) Intersecting arcs.
(b) Concentric circles.
(c) Trammel.
(d) Within a rectangle.

These methods will now be considered:

To construct an ellipse using intersecting arcs (Fig. 3.13c)

Draw the axes and focal points described previously.

Divide F1 F2 into an unequal number of divisions. Using compasses, set up A1 and strike an arc about F1. Reset compasses to 1C and strike an arc about F2 to intersect with the first arc. The point of intersection X of the arcs is a point on the ellipse since F1X + F2X = AC. Continuing along the major axis, arc A2 about F1 intersects arc 2C about F2 at Y. Arc A3 about F1 intersects arc 3C about F2 at Z and so on. The arcs may be struck in all four quadrants of the axes before resetting compasses for the next arc.

Although this method may be more laborious than the alternatives, it does prove conclusively the definition of the ellipse given earlier.

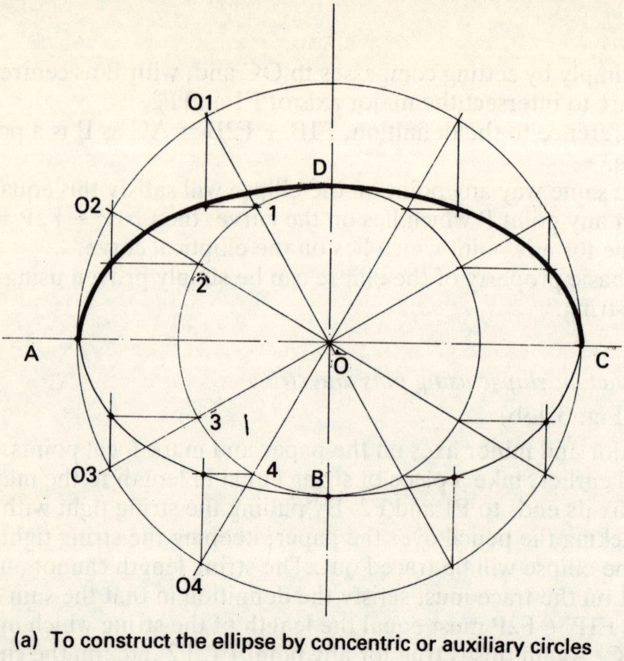

(a) To construct the ellipse by concentric or auxiliary circles

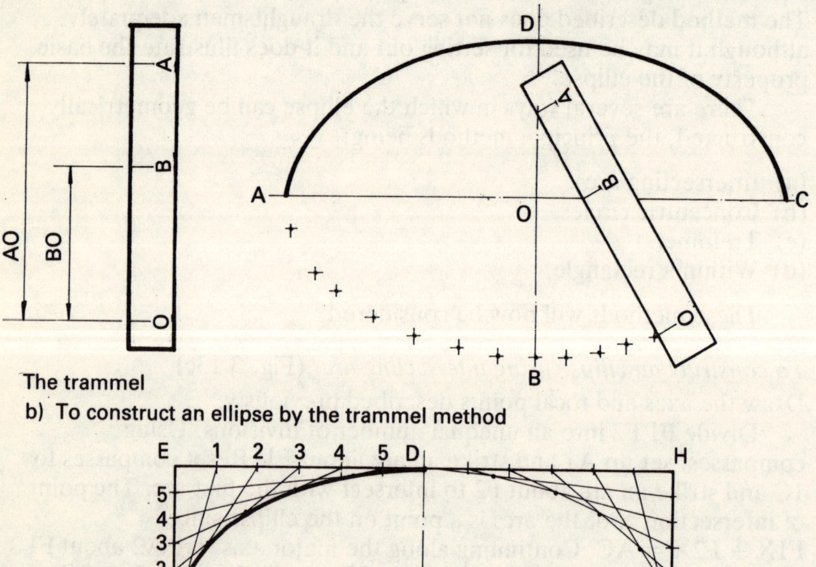

The trammel

b) To construct an ellipse by the trammel method

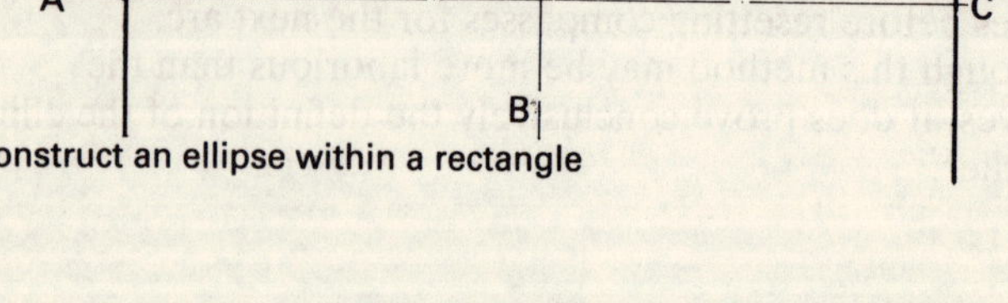

c) To construct an ellipse within a rectangle

Fig. 3.14

To construct an ellipse using concentric or auxiliary circles (Fig. 3.14a)

1. Set out given major axis and minor axis AC and BD.
2. With centre O construct two concentric circles of radius OA and OB.
3. From centre O extend any convenient number of radii to intersect the outer circle.
4. From the intersections $O1^1$, $O2^1$, $O3^1$, etc. drop perpendiculars to intersect horizontals from intersections 1, 2, 3, etc.
5. The intersections formed by the two projected lines are points on the elliptical curve; the points can now be joined with a smooth curve. Using this method it is not necessary to plot the focal points to draw the ellipse.

The trammel method (Fig. 3.14b)

1. Set out given axes as before.
2. On a suitable straight edged strip of paper (the trammel), mark off the lengths AO and BO with O being a common point to both.
3. Place the trammel onto the axes such that point A is on the minor axis and point B on the major. Mark off point O and move the trammel round keeping A and B on their respective axes, and marking the position of point O. The points made by O are points on the ellipse which can now be joined with a smooth curve.

Within a rectangle (Fig. 3.14c)

1. Construct a rectangle EFGH about the given axes of the ellipse to pass through points ABC and D.
2. Divide AE and ED into the same number of equal divisions.
3. Join A:1, 1:2, 2:3 and 3:4, giving points of intersection on the ellipse curve, which may be joined to form the curve.

Approximation to the ellipse using compass arcs (Fig. 3.15a)

1. Construct the major axis AC and the minor axis BD intersecting at O.
2. Divide OB into three equal parts at 1 and 2.
3. Mark off AE equal to O2 and CF also equal to O2.
4. Join E1 and construct its perpendicular bisector intersecting OD at G.
5. Construct circles of radius AE and FC about centres E and F.
6. From centre G construct GH passing through E.
7. About centre G draw arc of radius GH meeting circles at H and J.
8. Transfer centre G to OB produced and repeat arc to complete the ellipse.

It must be understood that the figure so drawn is not a true ellipse, but an approximation to it, all curves being part circular curves. In view of this the student should not use this method when required to produce a true ellipse.

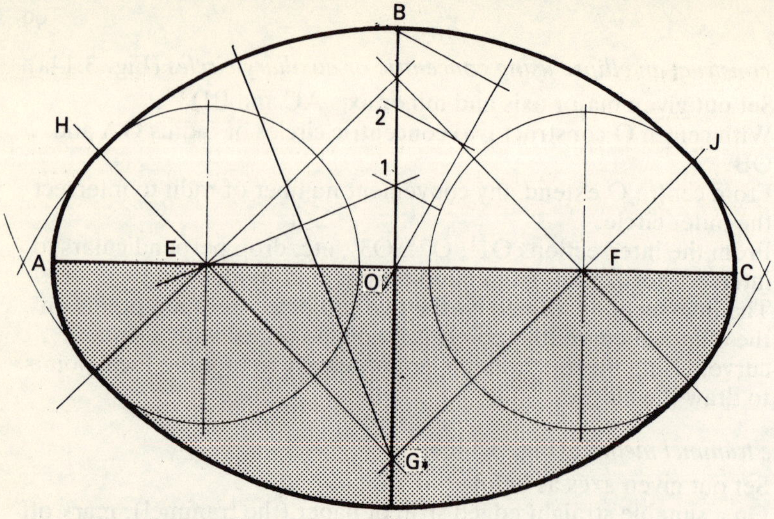

(a) To construct an approximate ellipse

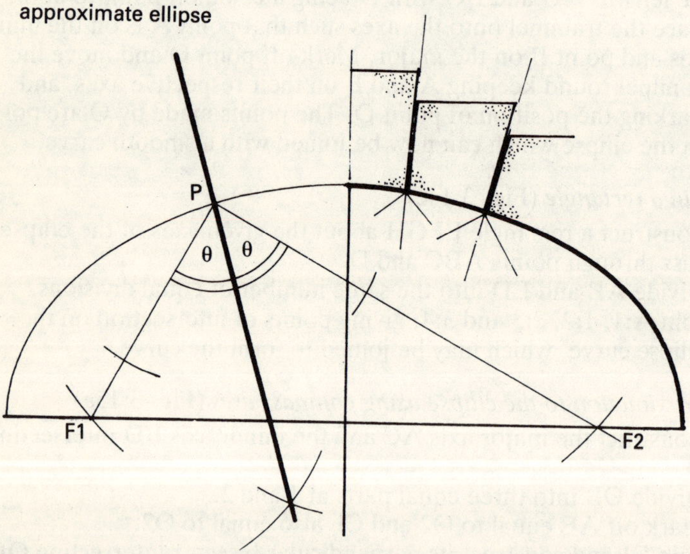

(b) To construct the normal to an eclipse

Fig. 3.15

To construct the normal to an ellipse (Fig. 3.15b)

1. Select the required point P on the ellipse.
2. Join the focal points F1 and F2 to the point P.
3. Bisect the angle F1 P F2 to give the required normal.

A practical application of this construction is shown in the setting out of a masonary arch.

The parabola

As well as being a conic section (Ch. 5), the parabola may be considered as the locus of a point. It may be defined as: *The path of a point which moves in such a way that its distance from a fixed point on the axis, called the focus, is equal to its perpendicular distance from a straight line called the directrix.*

The construction of the parabola

To construct a parabola given the directrix and the focus on the perpendicular (Fig. 3.16a)

1. Bisect FC to give E. The curve will pass through the point E, since FE = EC
2. Divide ED into horizontal divisions (which need not be equal), 1, 2, 3, etc.
3. With radius C1 and centre F draw arcs to cut horizontal 1 at 1′ and 1″.
4. Repeat for 2, 3, etc, using F as centre.
5. Join intersections to form the parabola.

To construct a parabola within a given rectangle ABCD (Fig. 3.16b)

1. Divide DE into a number of equal parts.
2. Divide AD into the same number of parts.

(By proportional division if necessary.)

3. Draw radial lines from points on AD to join at G.
4. Project vertical lines from the points on DE.
5. Join the intersections to form the parabola.

The hyperbola

The hyperbola may also be considered as the focus of a point such that it is: *The path of a point which moves such that its distance from the focus bears a constant ratio to its perpendicular distance from the directrix.* The ratio is always greater than unity (1 : 1).

To construct a hyperbola having a ratio of 1 : 2 and a focal distance of 30 mm (Fig. 3.17)

1. Construct the directrix AB and the axis CD.
2. Position the focal point F 30 mm from the directrix.
3. Divide the focal distance in the ratio 1 : 2 (by proportional division if necessary), index the points a, b and f (the curve will pass through point a).
4. Divide the remaining axis into a number of equal parts, and index them 1, 2, 3, 4, etc.

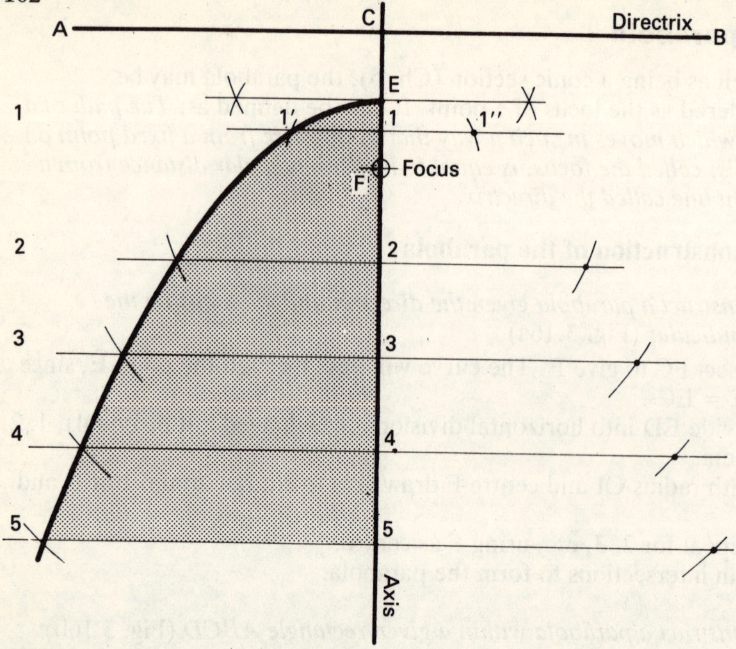

(a) The parabola as the locus of a point

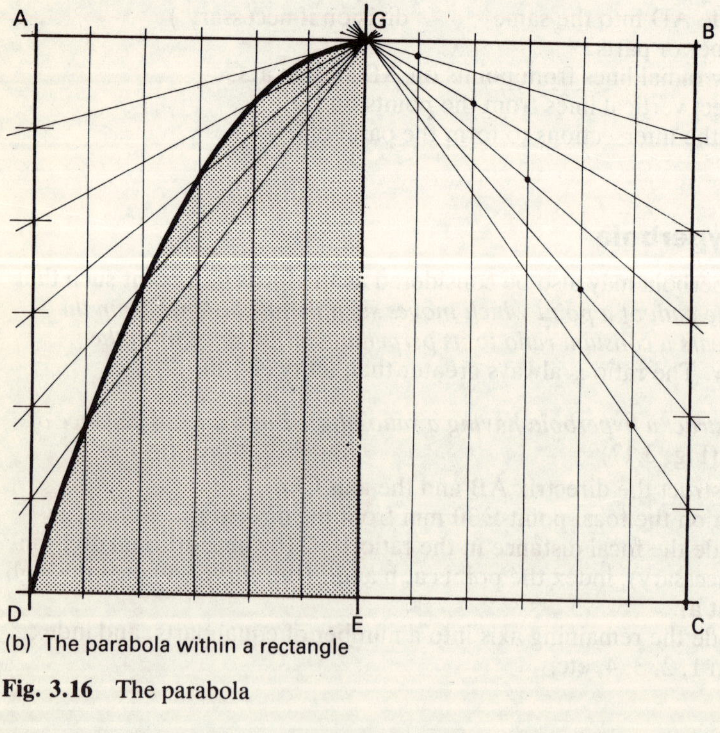

(b) The parabola within a rectangle

Fig. 3.16 The parabola

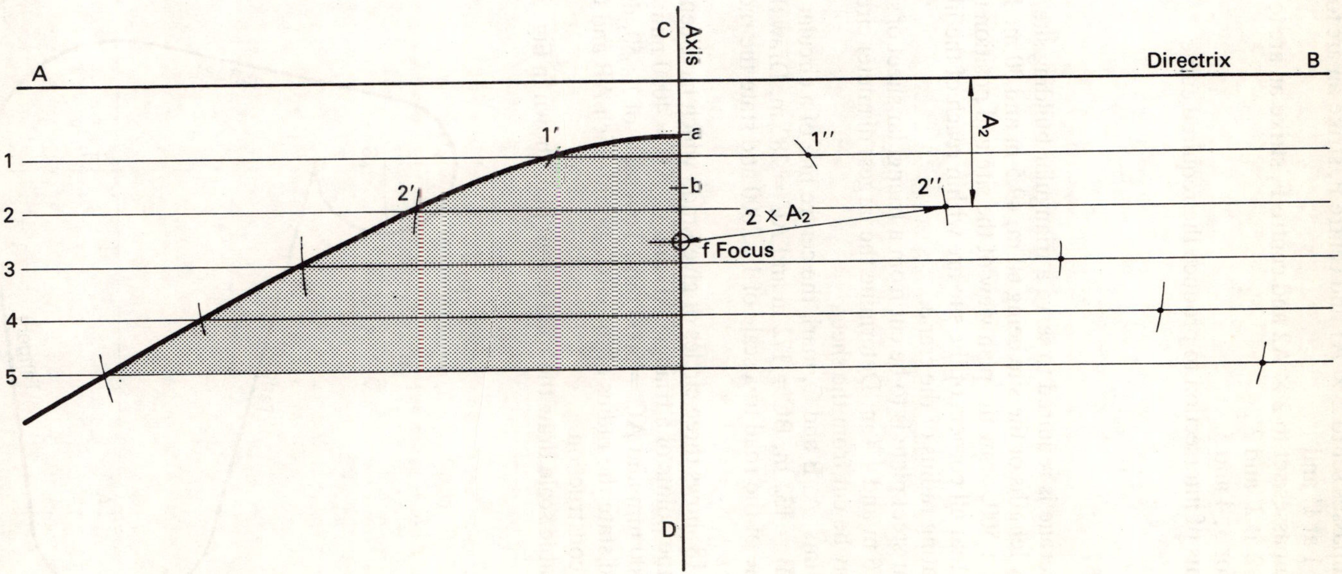

Fig. 3.17 The hyperbola

5. With compasses set to 2 × A1 and centre F, strike an arc to intersect 1 at 1′ and 1″.
6. With compasses set to 2 × A2 and centre F, strike an arc to intersect 2 at 2′ and 2″.
7. Repeat for 3, 4 and 5.
8. Join points of intersection to produce the required curve.

Exercises

1. A tower crane is required to serve a triangular building site, the boundary lengths of the site being 60 m, 49.5 m and 30 m. To a scale of 1 : 500, draw the plan view of the site and position the crane so that all corners of the site are within reach of the jib. State the operating radius of the crane.
2. A circular steel plate is to be cut from a triangular sheet of sides 6.0 m, 6.6 m and 4.4 m. Determine the largest diameter circle which may be cut from the sheet.
3. Three points, A, B and C, mark the centre line of a circular road curve, AB = 132 m, BC = 172 m and AC = 282 m. Draw the centre line of the road to a scale of 1 : 2000 and state the radius of the curve.
4. Figure 3.18 shows three circles of given radii which have centres lying on the points of a triangle AB and C. AB = 4000 mm BC = 3000 mm and AC = 5000 mm. To a scale of 1 : 50, draw the figure and state the radius of the circle upon which AB and C lie. Show all construction.
5. To a suitable scale draw the road layout plan shown in Fig. 3.19.

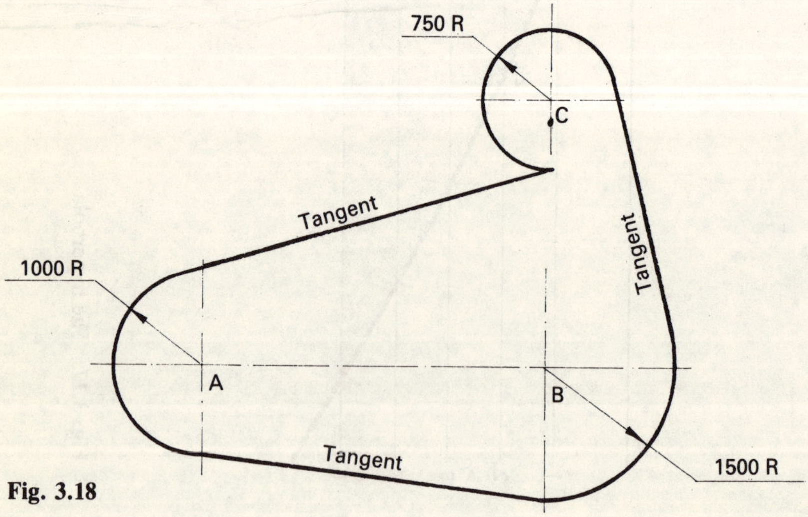

Fig. 3.18

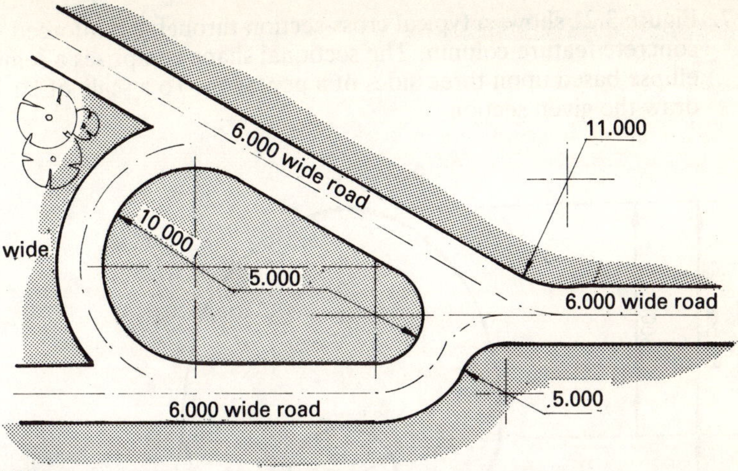

Fig. 3.19 PLAN

6. A sketch plan of an elliptical planter is shown in Fig. 3.20. The planter is capped with a 300 mm-wide artificial stone coping, bedded and jointed as shown on the sketch. To a scale of 1 : 20, draw and dimension the given plan showing the joints and all the necessary construction.

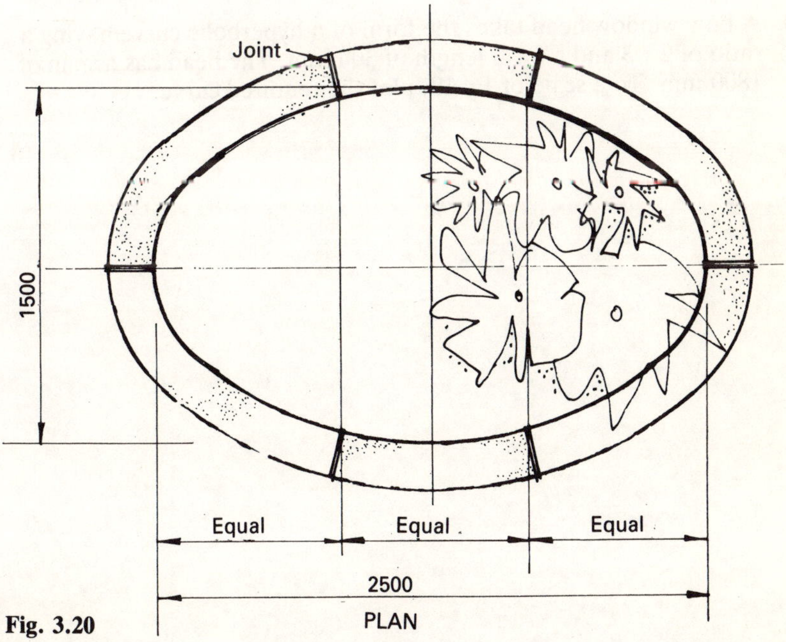

Fig. 3.20 PLAN

7. Figure 3.21 shows a typical cross-section through a reinforced concrete feature column. The sectional shape comprises a semi ellipse based upon three sides of a pentagon. To a scale of 1 : 10, draw the given section.

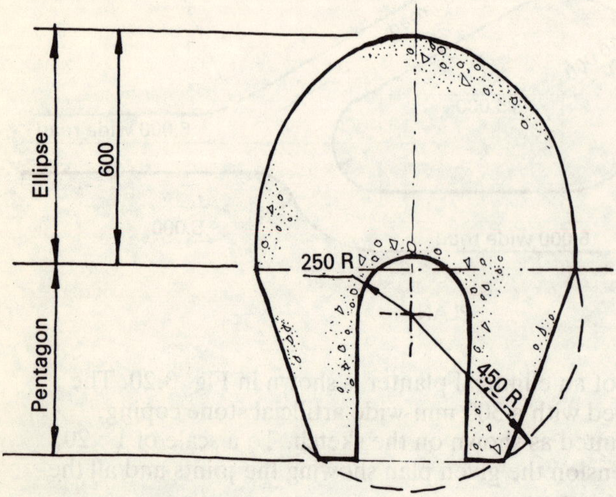

Fig. 3.21 Cross-section

8. A bow window head takes the form of a hyperbolic curve having a ratio of 2 : 3 and a focal length of 300 mm. The head has a span of 1800 mm. To a scale of 1 : 10, plot the required curve.

Chapter 4

Drawing to scale

Because of the nature of building and civil engineering works, it is often impractical, and indeed unnecessary, to produce full-size 1 : 1 drawings. In general, drawings of buildings, surveys, assemblies and components are drawn in proportion. It is important to realize then that, although the plans and elevations of say a multi-storey building may be produced within the confines of a sheet of paper, the relative proportions of height, width and thickness are accurately represented. This is the essence of scale drawing: the accurate representation of an object usually at a ratio of less than 1 : 1 (full size), and in so doing preserving the correct relationship of its three-dimensional properties.

It is now generally accepted practice in the construction industry that the metric system of units is used in the drawing process, and it is assumed that the student is familiar with metric linear measurement. Reference will be made to imperial scales, however, in view of the fact that some drawings based on imperial measurement will still be in circulation.

It is essential that the student has a firm understanding of scale, together with a competent practical application of scale rules, bearing in mind that virtually all of the work he will produce will be drawn to some diminutive scale or another.

A most useful exercise in gaining an understanding of scale is in the construction of scales on the drawing sheet. Although this may be rather an academic exercise – as under normal circumstances standard scales and scale rules will be used – it is certainly possible that a practical situation may arise which necessitates the construction of an obscure, or even unique, scale. In practice scales such as $\frac{1}{4}$ inch to 1 metre are not unheard of; needless to say, this is ridiculous.

First consider a few of the scales commonly used in practice; for instance, 1 : 5. This ratio means that 1 mm on the drawing represents 5 mm actual. This may be considered the same as $\frac{1}{5}$ full size; the resulting fraction is known as the representative fraction. Furthermore, the scale may be thought of as 200 mm represents 1 metre.

All of these scales depicted in various forms are then the same:

1 : 5
$\frac{1}{5}$ full size
or 200 mm = 1 m.

Similarily, a scale of 1 : 200 may be depicted as

$\frac{1}{200}$ full size
or 5 mm = 1 m

and a scale of 1 : 100 may be
$\frac{1}{100}$ full size
or 10 mm = 1 m

Linear scales

Linear scales can be produced simply by a process of proportional division as described in Chapter 3.

Example 1 To construct a linear scale of 1 : 20 (Fig. 4.1a)

1. Draw the base line AB of any suitable length, remembering that the drawn line length will represent an actual distance, to a scale of 1 : 20. If AB is 200 mm long it will represent an actual distance of 200 mm × 20 = 4000 mm = 4.000 m.
2. Divide AB into four divisions, each representing 1.000 m.
3. Divide AC into 10 equal divisions. Each division will represent 0.1 m (100 mm).
4. Indicate numerical values on the scale, making point C zero, thus allowing the sub-divided scale to extend beyond the zero point. This assists in the practical use of the scale, when the divider points can be correctly positioned without making adjustments on the scale for the sub-divisions.
5. Mark off the 1 m values along the base line as shown in the figure. The sub-divided scale can now be marked off with values of 0.1 m.

Provided that the scale is drawn accurately scale measurements up to 4.000 m can be made to an accuracy of 0.1. A distance of 3.600 m is indicated on the scale and this can be verified with a suitable scale rule.

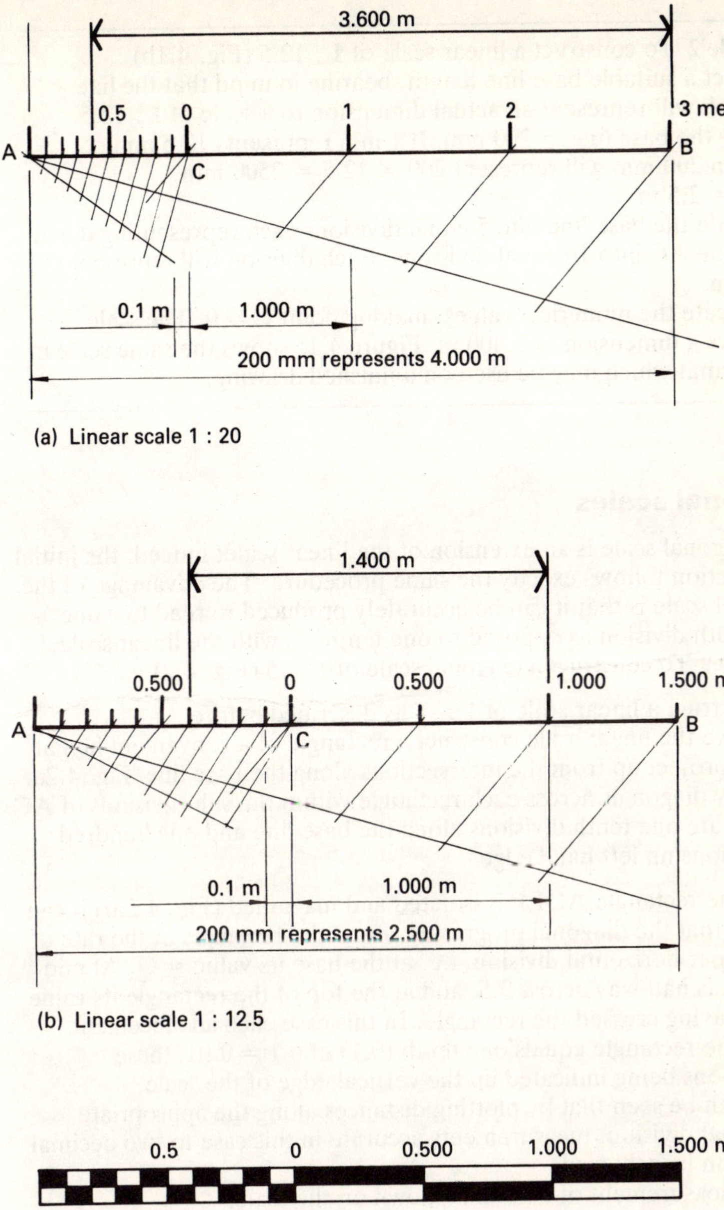

(a) Linear scale 1 : 20

(b) Linear scale 1 : 12.5

(c) 1 : 12.5 scale

Fig. 4.1

Example 2 To construct a linear scale of 1 : 12.5 (Fig. 4.1b)

1. Select a suitable base line length, bearing in mind that the line length will represent an actual dimension to a scale of 1 : 12.5.
   Say the base line = 200 mm. If 1 mm represents 12.5 mm then 200 mm will represent 200 × 12.5 = 2500 mm,
which = 2.5 m
2. Divide the base line into 5 equal divisions each representing 0.5 m.
3. Divide AC into 10 equal divisions. Each division will represent 0.1 m.
4. Indicate the numerical values, making point C = 0. The scale shows a dimension of 1.400 m. Figure 4.1c shows the same scale in a format which may be used on a finished drawing.

Diagonal scales

The diagonal scale is an extension of the linear scale; indeed, the initial construction follows exactly the same procedure. The advantage of the diagonal scale is that it can be accurately produced to read to a one hundredth division as opposed to one tenth, as with the linear scale.

Example: To construct a diagonal scale of 1 : 15 (Fig. 4.2)

1. Construct a linear scale of 1 : 15 as described before.
2. Above the linear scale construct a rectangle of a convenient height and project up from the intersections along the base line. Fig. 4.2a.
3. Draw diagonals across each rectangle within the sub-divisions of AC.
4. Indicate one tenth divisions along the base line and one hundred divisions up left hand edge.

If the rectangle ADEF is isolated and magnified (Fig. 4.2.b) it can be seen that the diagonal progresses across the rectangle at the rate of $\frac{1}{10}$(0.1) per horizontal division, i.e. at the base its value = O. At mid height it is half way across 0.5, and at the top of the rectangle its value is 1.0, having crossed the rectangle. In this case each advancement across the rectangle equals one tenth (0.1) of 0.1 = 0.01, these graduations being indicated up the vertical edge of the scale.

It can be seen that by plotting distances along the appropriate horizontal division, measurements accurate in this case to two decimal places can be achieved.

A measurement of 2.46 m is shown on the scale.

The selection of scales

In practice, unusual scales should be avoided as this can lead to misunderstanding and frustration. Table 4.1 gives a range of preferred scales for various types of drawing.

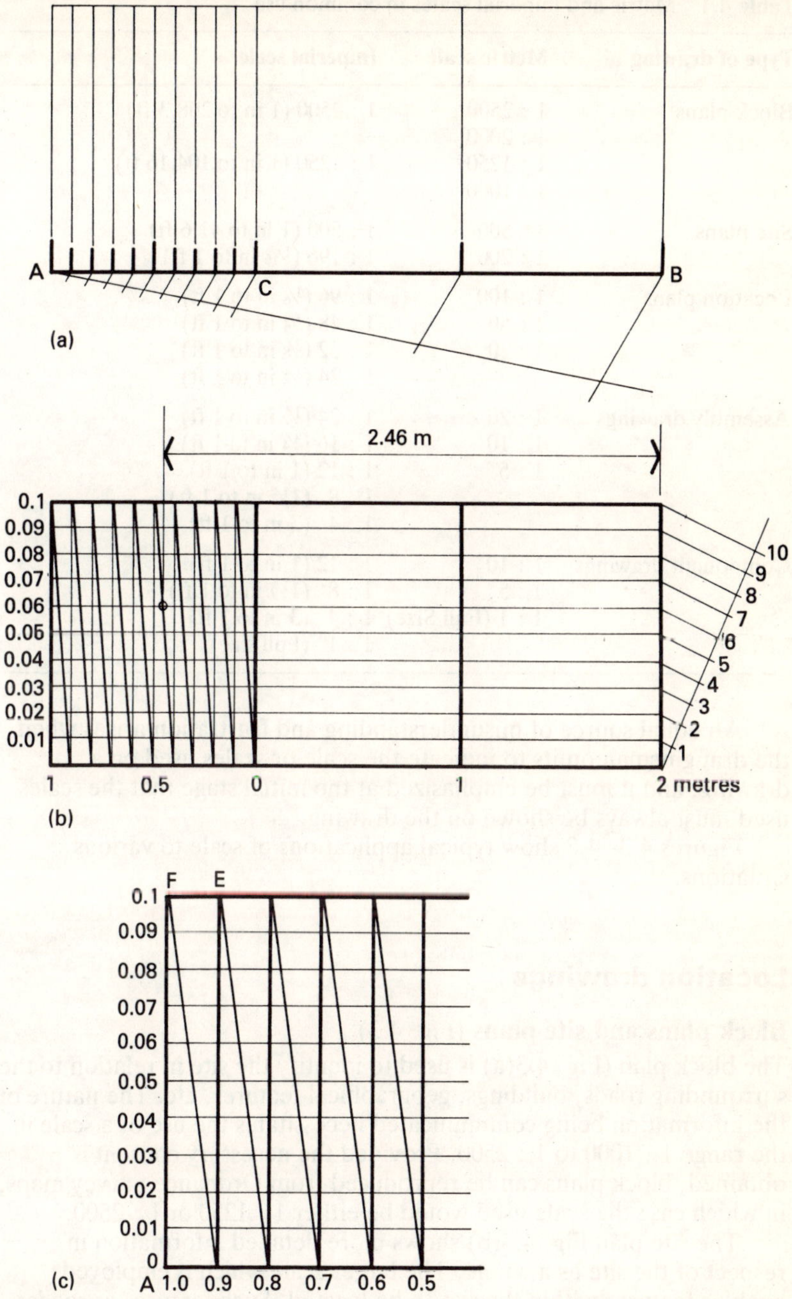

Fig. 4.2 To construct a diagonal scale of 1:15

Table 4.1 Metric and imperial scales in common use

Type of drawing	Metric scale	Imperial scale
Block plans	1 : 2500	1 : 2500 (1 in to 208.3 ft)
	1 : 2000	
	1 : 1250	1 : 1250 (1 in to 104.16 ft)
	1 : 1000	
Site plans	1 : 500	1 : 500 (1 in to 41.6 ft)
	1 : 200	1 : 196 (1⁄16 in to 1 ft)
Location plans	1 : 100	1 : 96 (⅛ in to 1 ft)
	1 : 50	1 : 48 (¼ in to 1 ft)
	1 : 20	1 : 32 (⅜ in to 1 ft)
		1 : 24 (½ in to 1 ft)
Assembly drawings	1 : 20	1 : 24 (½ in to 1 ft)
	1 : 10	1 : 16 (¾ in to 1 ft)
	1 : 5	1 : 12 (1 in to 1 ft)
		1 : 8 (1½ in to 1 ft)
		1 : 4 (3 in to 1 ft)
Component drawings	1 : 10	1 : 12 (1 in to 1 ft)
	1 : 5	1 : 8 (1½ in to 1 ft)
	1 : 1 (Full Size)	1 : 4 (3 in to 1 ft)
		1 : 1 (Full size)

An equal source of misunderstanding and frustration arises when the draughtsman omits to indicate the scale or scales used on a drawing, and it must be emphasized at the initial stage that the scales used must always be shown on the drawing.

Figures 4.3–4.8 show typical applications of scale to various situations.

Location drawings

Block plans and site plans (Fig. 4.3)

The block plan (Fig. 4.3(a) is used to identify the site in relation to the surrounding roads, buildings, geographical features, etc. The nature of the information being communicated necessitates the use of a scale in the range 1 : 1000 to 1 : 2500. Provided the necessary consent is obtained, block plans can be reproduced from Ordnance Survey maps, in which case the scale used would be either 1 : 1250 or 1 : 2500.

The site plan Fig. 4.3(b) shows more detailed information in respect of the site as a whole. The larger scale which is employed enables features within the site to be located. Such features as roads, footways, trees, walls and outbuildings can be indicated; the building can be positioned on the site, its position being related to a base line or

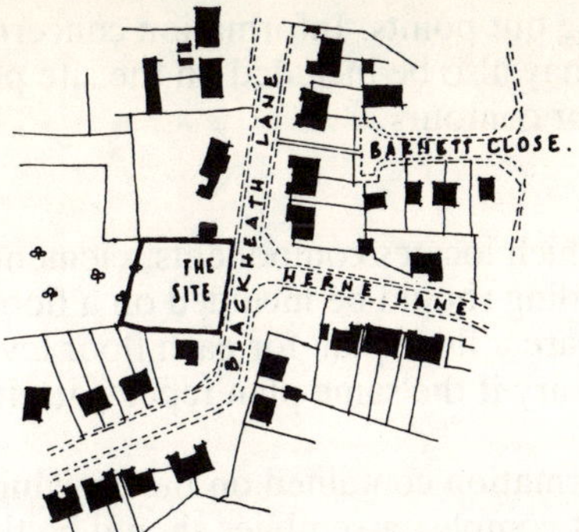

(a) BLOCK PLAN. Scale 1: 2500.

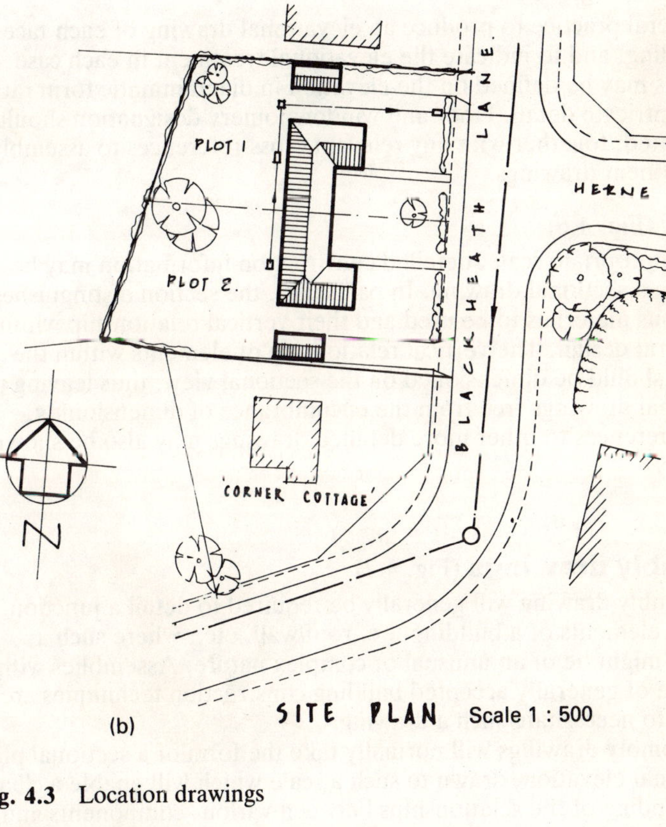

(b) SITE PLAN Scale 1 : 500

Fig. 4.3 Location drawings

to distinguishable setting out points. Information concerning the topography of the site may also be included on the site plan, either in the form of spot levels or contours.

Floor plans (Fig. 4.4)

Selective information which locates components, elements and assemblies within a building should be included on a floor plan. It is general practice to prepare a floor plan for each floor level although this may prove unnecessary if the same plan repeats itself several times.

The amount of information contained on the drawing will depend upon the scale used, but complex assemblies should be the subject of a separate drawing which may be cross-referenced on the floor plan, this in turn safeguards against overcrowding.

Elevations (Fig. 4.5)

It is general practice to produce an elevational drawing of each face of the building, and to indicate the elevational treatment in each case. Materials may be defined on the elevation in diagrammatic form rather than in intricate detail. Door and window joinery designation should be indicated, together with any relevant cross-references to assembly or component drawings.

Sections (Fig. 4.6)

At the appropriate scale, detailed construction information may be shown on a sectional drawing. In particular, the section distinguishes the various materials to be used and their vertical relationship within the general design. The vertical relationship of elements within the building should be dimensioned on the sectional view, thus leaving the elevational drawings free from the encumbrance of dimensioning. Cross-references to other more detailed drawings may also be shown if necessary.

Assembly drawings (Fig. 4.7)

An assembly drawing will generally be required to detail a junction between elements of a building, i.e. roof/wall, etc., where such a junction might be of an unusual or complex nature. Assemblies within the scope of generally accepted building construction techniques are unlikely to necessitate such a drawing.

Assembly drawings will normally take the form of a sectional plan or sectional elevation, drawn to such a scale which will enable a clear understanding of the relationships between various components and materials. Standard assembly details such as that illustrated in Fig. 4.7 may be used repetitively.

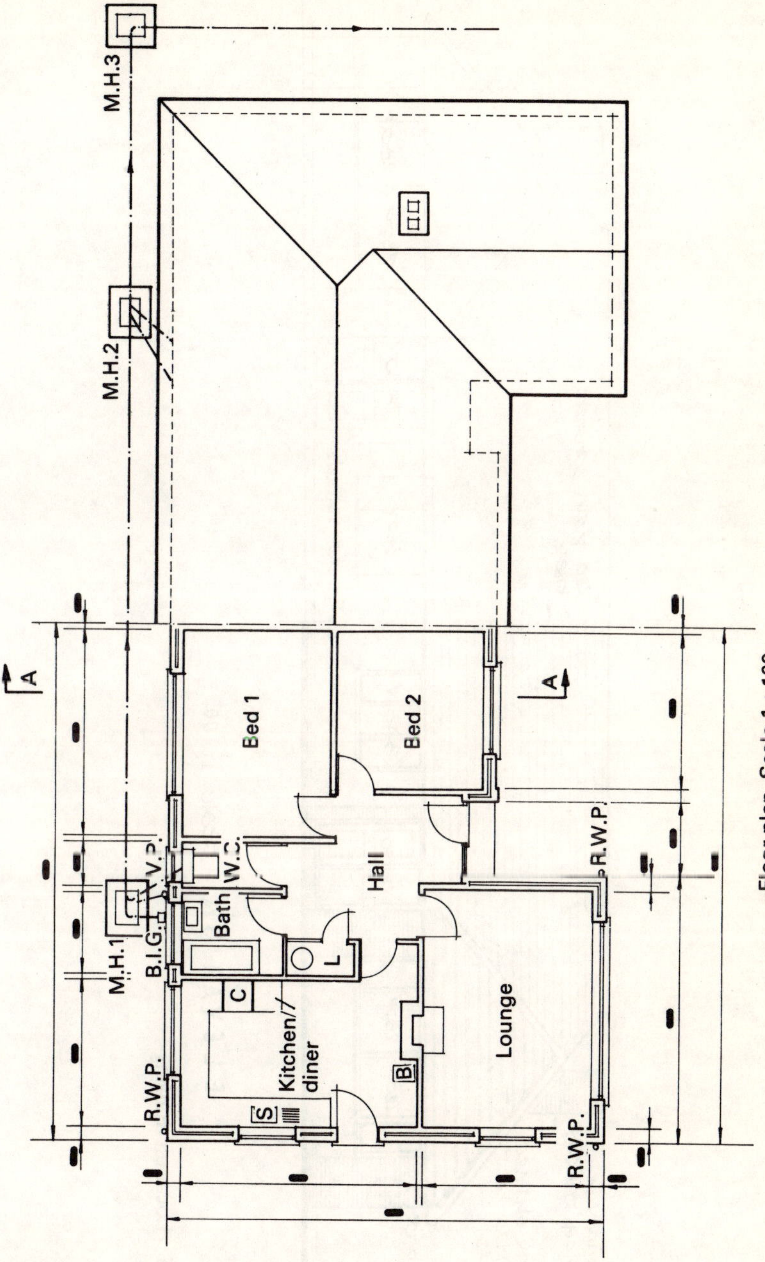

Fig. 4.4 Location drawing

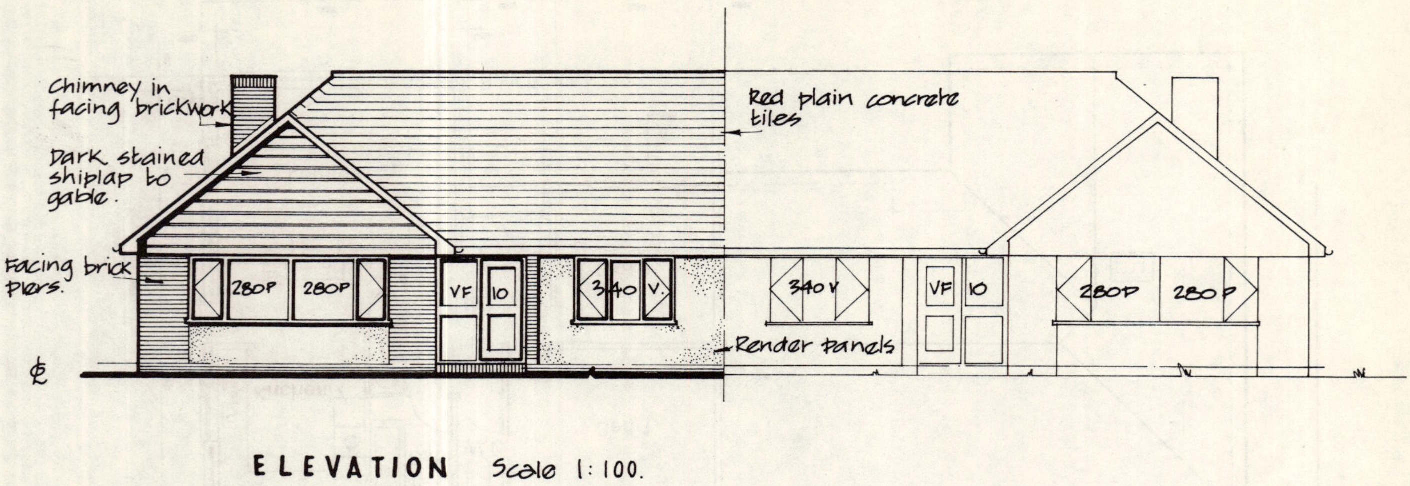

Fig. 4.5 Location drawing

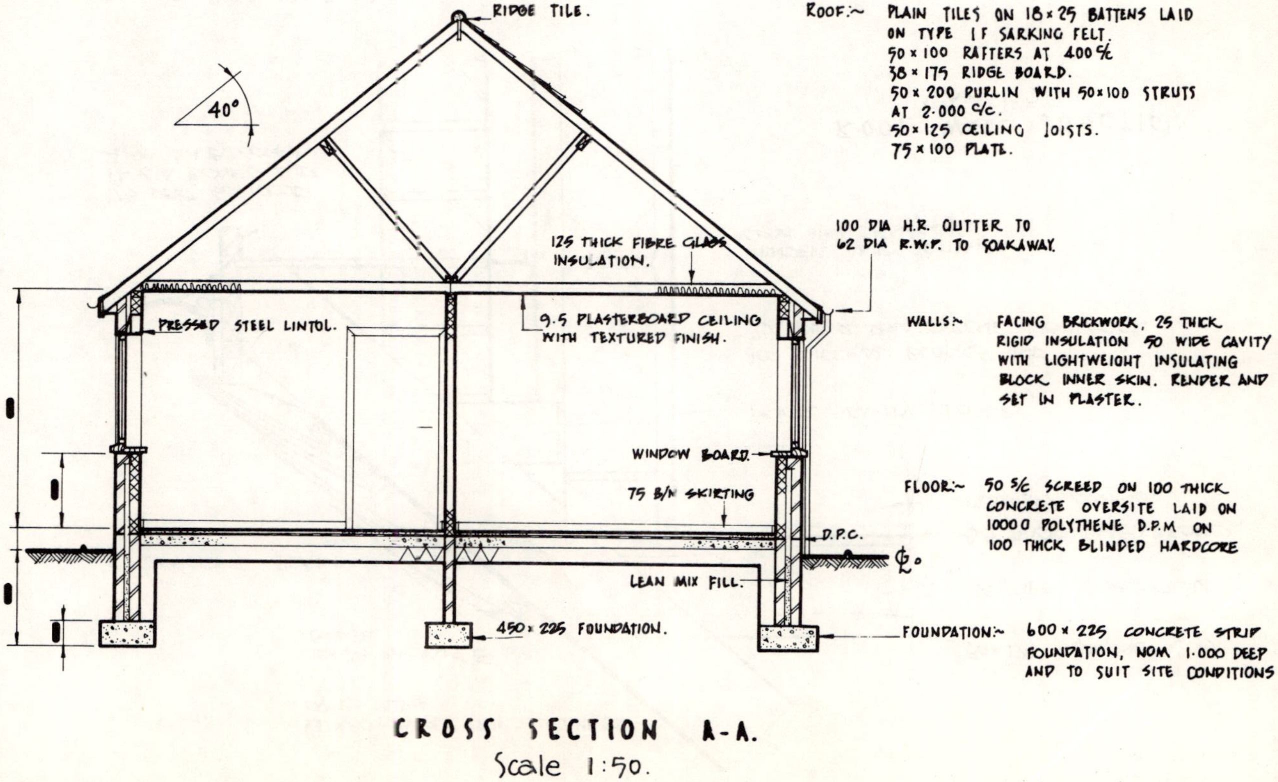

Fig. 4.6 Location drawing

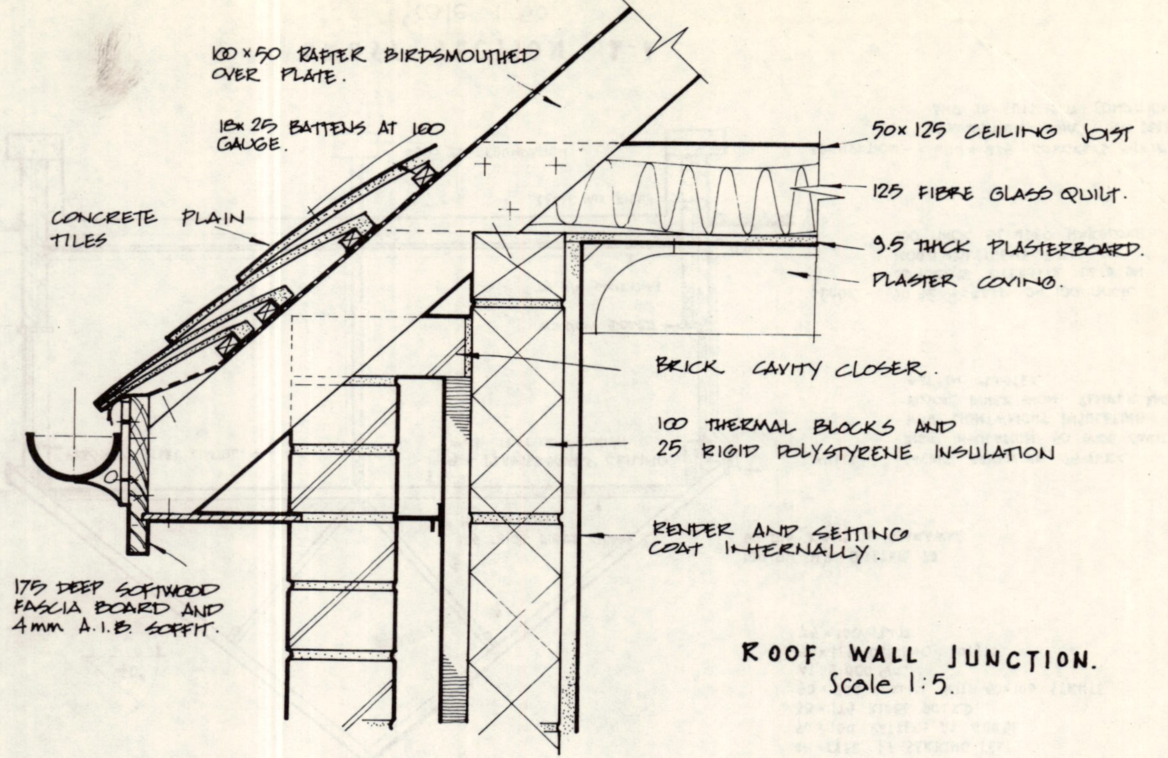

Fig. 4.7 Assembly drawing

Component drawings (Fig. 4.8)

Parts of an assembly which require further detailing may be isolated and made the subject of a component drawing. The principal purpose of this type of drawing is to assist in the manufacture of a specific component, when insufficient detail is included on the assembly drawing.

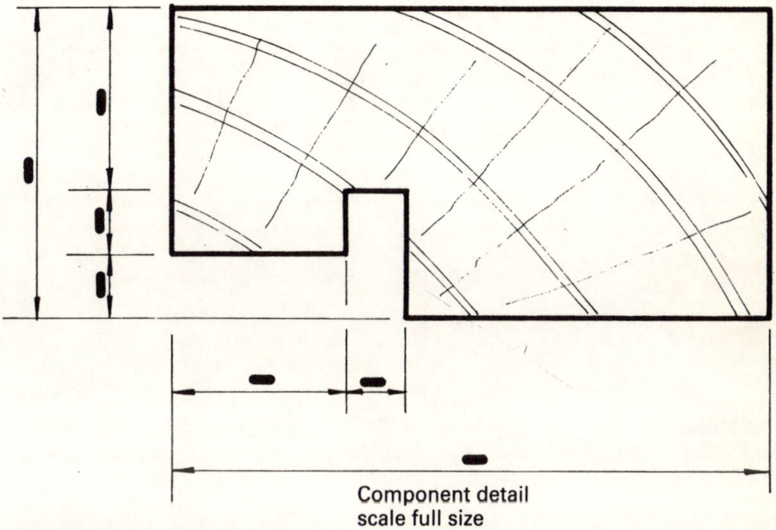

Fig. 4.8 Component drawing

Exercises. Drawing to scale

1. On a base line 150 mm long, construct a scale of 1 : 20 to read to an accuracy of 0.1 m. With the aid of a scale rule, check the accuracy of the drawn scale.
2. A line 200 mm long represents 50 km.
   (a) Calculate the scale.
   (b) Construct a scale upon this base line to read to an accuracy of 1 km.
   (c) Indicate on the scale a distance of 34 km.
3. A site measurement of 20 m scales 62.5 mm on a plan.
   (a) Calculate the scale of the plan.
   (b) Construct a suitable scale for use with the plan.
   (c) Indicate on the scale a distance of 58 km.
4. Two points 275 mm apart on a map are in fact 5.5 km distant.
   (a) Calculate the scale of the map.

(b) Construct a suitable scale for use with the map, capable of measuring to an accuracy of 0.01 km.

5. Select a suitable scale, and prepare an assembly drawing for the sub-structure of the bungalow shown in Fig. 4.6. Fully annotate and dimension the drawing.

Chapter 5

Projections

Types of projection

It is an essential of any drawing to accurately convey the three-dimensional properties of the object to be illustrated.

The draughtsman's task is to represent these three dimensions of length, height and thickness on a two-dimensional surface in the form of the drawing sheet.

To achieve uniformity in representation, there are recognized methods of approach to the drawing and setting out of views known generically as projections. They may be considered under three headings, these being:

- Multi-plane
- Axonometric
- Oblique

These may be further sub-divided as shown in Fig. 5.1, extracted from BS 1192.

Multi-plane projection

As the name suggests, multi-plane projection is a method of three-dimensional representation making use of 'many' planes upon which the object to be drawn can be projected.

The planes described are known as planes of projection, and are, for all practical purposes, imaginary.

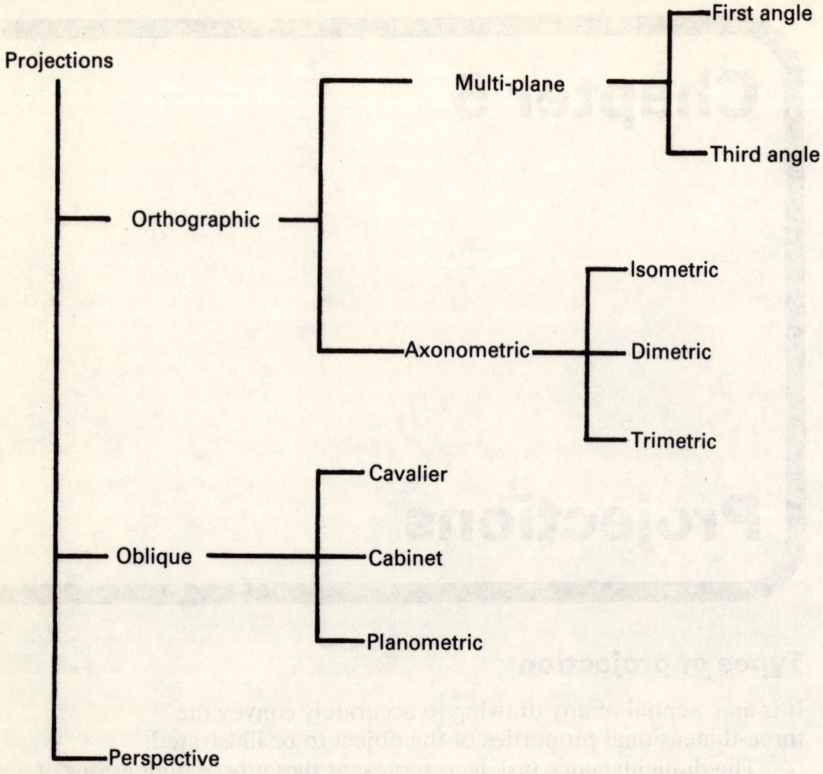

Fig. 5.1

The common approach to projection onto planes is based upon a recognized arrangement of the principal planes in space, this arrangement being illustrated in Fig. 5.2.

By reference to Fig. 5.2 it can be seen that the principal planes take the form of three flat surfaces, each intersecting or abutting the other at right angles.

The principal planes shown are known as:

- The vertical plane
- The horizontal plane
- The side vertical plane

The cruciform format of the planes divides them into four easily identifiable quadrants known as the first, second, third and fourth quadrants. The second and fourth quadrants may be disregarded as they have no function in projection.

The first and third, however, are all important to the draughtsman as all multi-plane projection will be made in one or other of these quadrants.

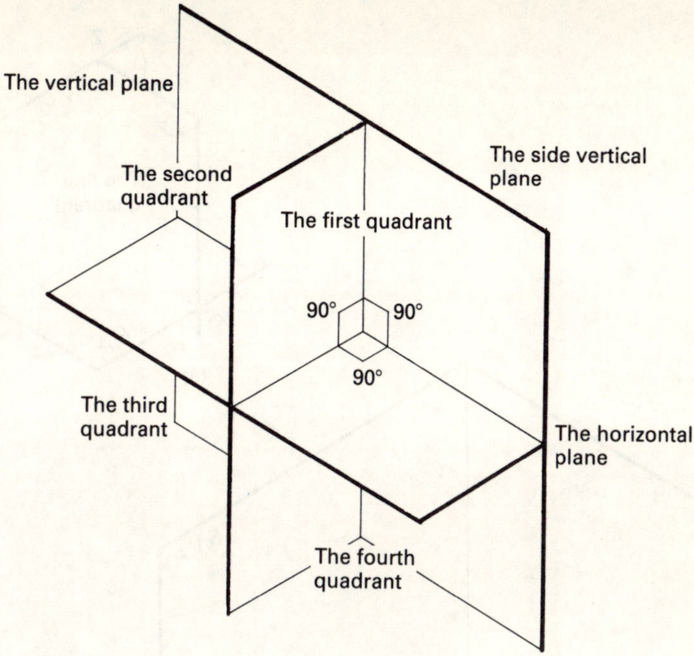

Fig. 5.2 The principal planes of projection

First angle projection

The term first angle describes projection which is carried out in the first quadrant. It is generally accepted in the construction industry that first angle projection is used in the production of drawings.

The three principal planes in the first quadrant are isolated in Fig. 5.3a, the remaining three quadrants having been disregarded.

The three planes may be considered as the three sides of a box, which are hinged along the lines of the joints XY and YZ.

The application of the planes can now be considered. The object to be drawn is suspended in a fixed position within the quadrant with its principal faces parallel with the planes. The outline and the features of the object are now projected onto the planes by projectors. The projectors are imaginary lines which project the object onto the planes, and, most important, are always perpendicular to them.

Figure 5.3b shows a simple object in the first quadrant. It can be seen from the figure that each plane displays a particular view of the object as follows:

- The **vertical plane (V.P.)** shows height and length and is known as the **front elevation**.
- The **side vertical plane (S.V.P.)** shows height and thickness and is known as the **end elevation**.

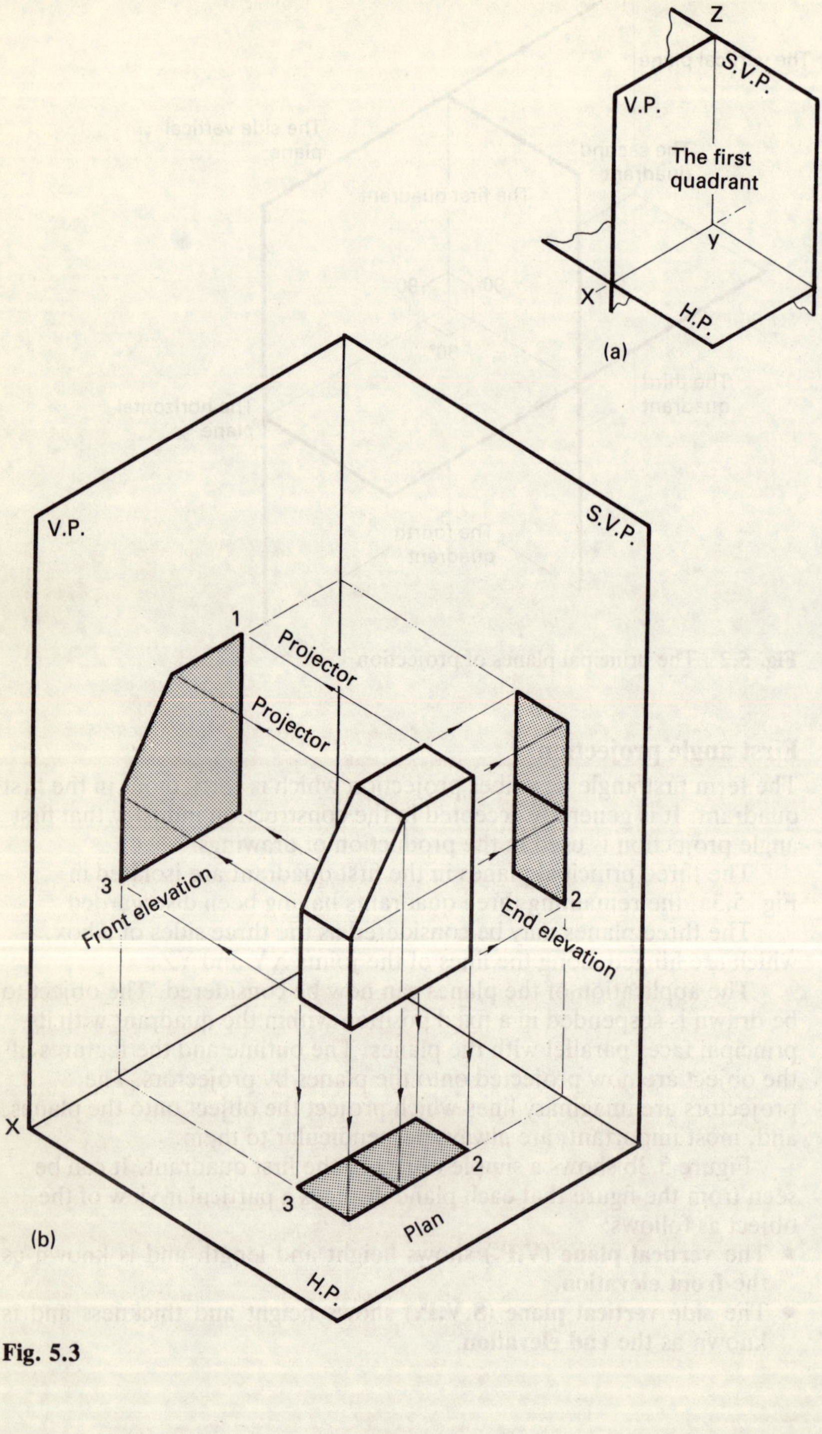
Z
S.V.P.
V.P.
The first quadrant
y
X
H.P.
(a)
V.P.
S.V.P.
1
Projector
Projector
3
Front elevation
2
End elevation
X
2
3
Plan
(b)
H.P.

Fig. 5.3

- The **horizontal plane (H.P.)** shows length and thickness and is known as the **plan**.

Having projected the object onto the three planes, the planes are now opened out onto a flat surface, rotating the H.P. about XY and the S.V.P. about YZ, bringing both into the V.P. The XY line has particular significance as will be seen later. The arrangement of the planes, and views resulting from this operation is shown in Fig. 5.4a. This arrangement is the accepted format for first angle projection and is recognized by the symbol in Fig. 5.4b.

In practice it is unnecessary to actually draw the planes of projection and after little experience the student should be able to visualize the appearance of views on the planes. To simplify the projection process a commonly used technique as shown in Fig. 5.4c can be adopted. The basis of this method can be appreciated by reference to Fig. 5.3b, where it can be seen that projectors share common vertical and horizontal planes within the quadrant as illustrated by the lines 1–1 2–2 and 3–3.

As a general rule for first angle projection it may be useful to remember that:

1. The view from the left appears on the right.
2. The view from above appears below.
3. The view from the right appears on the left.

Third angle projection

Third angle projection is little used for construction drawing in Great Britain, but its use overseas is more widespread. Certain manufacturing industries have adopted third angle as standard presentation, and for this reason the student should be able to recognize and prepare drawings on this format.

This method follows similar principles as described in first angle. The views produced are in fact identical, but, most important, the arrangement of the views and the way in which they are produced are fundamentally different.

Third angle projection is undertaken in the third quadrant, which is shown in Fig. 5.5a. The object to be drawn is suspended within the quadrant as previously described. It can readily be seen in Fig. 5.5b that in this situation the vertical plane of projection lies between the object and the viewer. This establishes the underlying principle of third angle projection in that all of the planes of projection are considered to be transparent and the subject is viewed through each of the planes in turn and then projected onto the respective plane.

If the planes are opened out about the axes XY and YZ1 into the vertical plane, as shown in Fig. 5.6a, it can be seen that the arrangement of the planes and consequently the views differ from first angle and it should be noted in particular that the plan view appears above the front elevation.

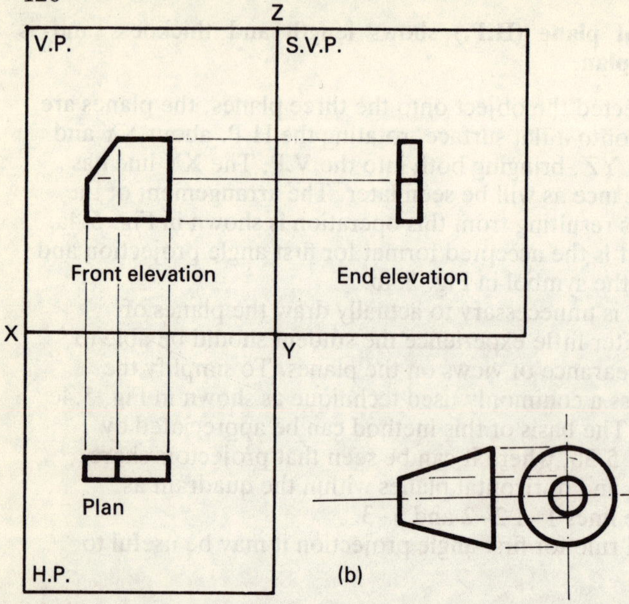

(a)

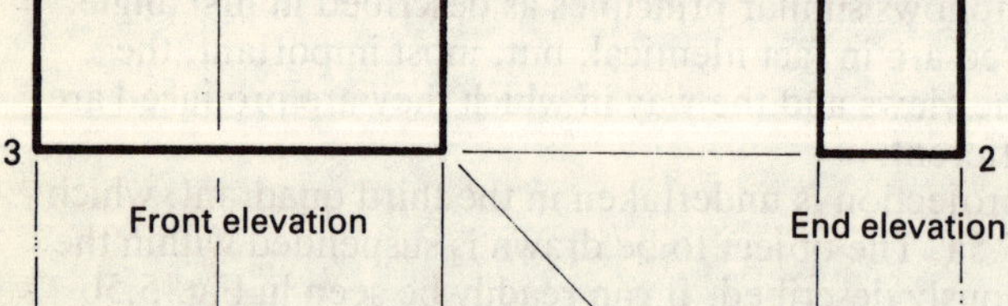

Fig. 5.4 First angle projection

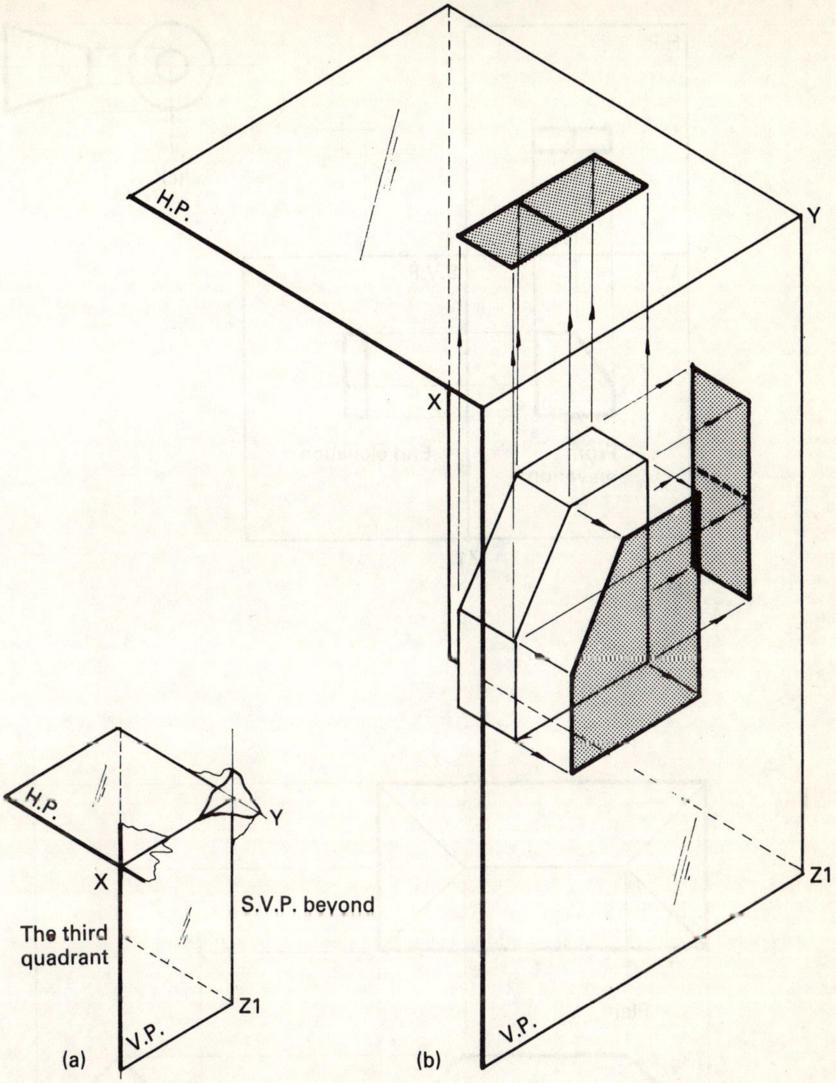

Fig. 5.5 Third angle projection

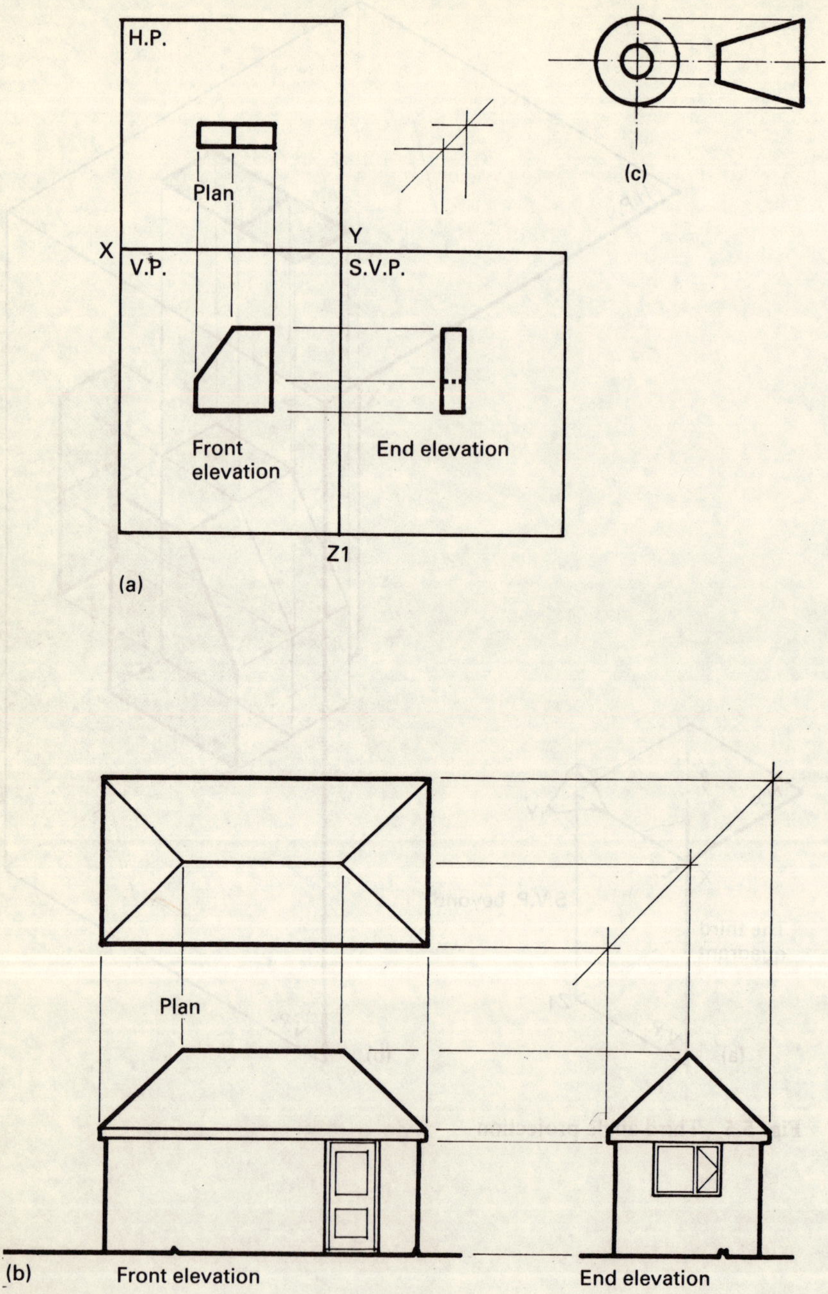
H.P.
Plan
X
Y
V.P.
S.V.P.
Front elevation
End elevation
Z1
(a)
(c)
Plan
(b)
Front elevation
End elevation

Fig. 5.6

It can be said that in third angle projection:

1. The view from the left appears on the left.
2. The view from the right appears on the right.
3. The view from above appears above.

Figure 5.6b shows the typical application of third angle projection to a simple building.

The recognized symbol for third angle projection is shown in Fig. 5.6c.

Hidden detail

When an object is illustrated in projection, invariably certain details and features of the object will not be visible, being obscured from view by the solidity of the object itself. To convey the maximum information, such obscured detail which is important to the understanding of the drawing may be shown in hidden detail using a medium, broken line (Fig. 2.2a, p. 45).

In construction drawing practice, the degree to which hidden detail is used is often a matter of discretion for the draughtsman. As an example, for component detailing and assembly drawings, hidden detail would generally be shown. On the other hand, for location plans, elevations, and similar drawings the use of hidden detail would be limited.

When showing hidden detail care must be taken with linework to ensure that the drawing is not dominated by lines which may be considered of secondary importance. When dealing with orthographic sections it will be noted that hidden detail is not shown beyond the cutting plane on the sectional view.

The drawing of hidden detail lines should be practised, bearing in mind the following basic rules.

1. Lines must be of consistent thickness with equal dash length and equidistant spacing (special pens are available for dotted lines in inkwork)
2. Points of juncture and extremities of dotted lines must be shown with a dash as shown in Fig. 2.4e (p. 48) and not separated by a space.

Attention to such detail distinguishes good practice from the mediocre.

Sections

Sectional drawing is of particular importance in construction drawing since the majority of information concerning a structure in terms of materials, structural frame, levels, secondary components and so on are shown on sectional plans and sectional elevations.

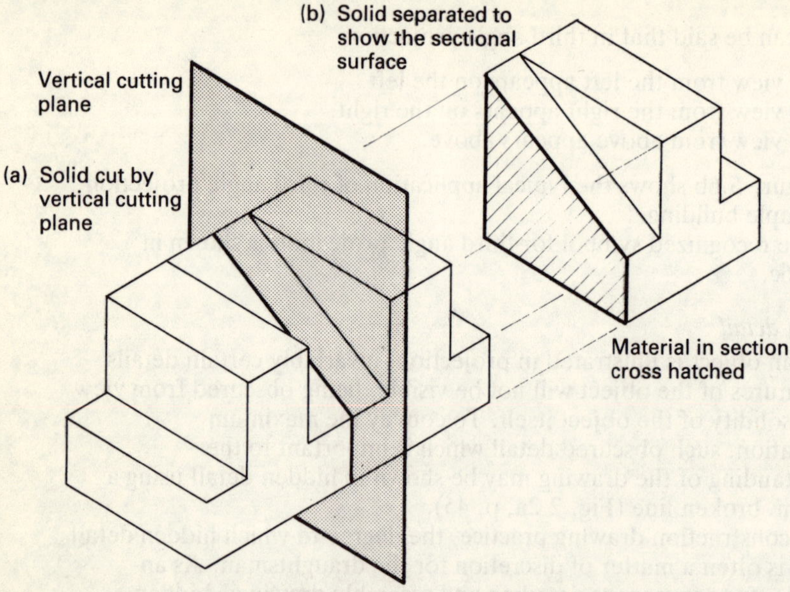

Fig. 5.7 Explanatory view of the vertical cutting plane

The function of sectioning is to show parts of an object which are obscured from view and would otherwise have to be shown in hidden detail.

The operation of sectioning is an imaginary process used to illustrate the view of an object when cut through by a cutting plane and that portion in front of the cutting plane is removed. Figure 5.7 illustrates this principle, where a vertical plane cuts through a timber block. Note that where the cutting plane has passed through the actual material the area is cross-hatched on the drawing. Further, the detail of the object beyond the cutting plane is shown although hidden detail has been omitted.

Several features shown on the drawing (Fig. 5.8) are of particular importance and need further mention.

In this particular case the cross-hatching is shown in the generally accepted manner, that is, thin lines, sloped at a constant 45° and spaced equidistant. The clarity of a sectional drawing depends upon consistency of line thickness slope and spacing and with little practice can be executed neatly and quickly. Although all three aspects contribute to clarity, spacing of lines should be given particular attention. It is invariably up to the draughtsman to determine spacing, but a common error is to place section lines too close together, causing an overpowering section; or alternatively too far apart, causing loss of clarity. As a rule it is necessary to consider scale, the amount and the nature of the hatching before the degree of sectioning can be decided.

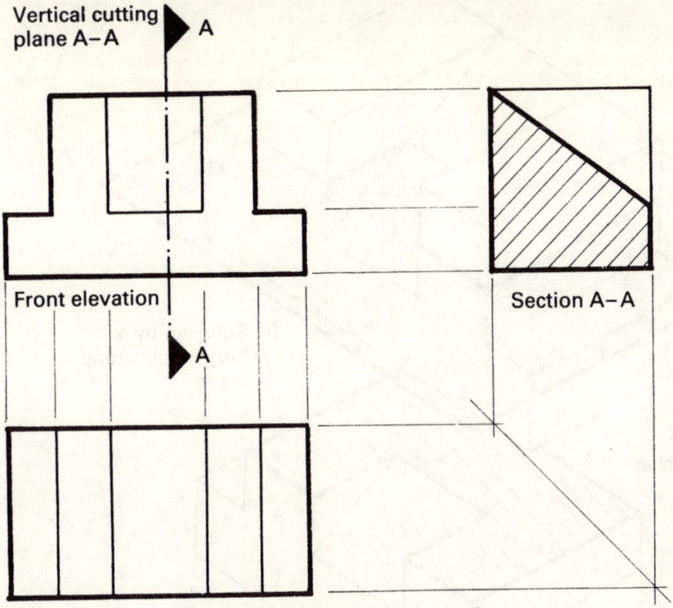

The sectional elevation shown in first angle projection

Fig. 5.8

This is particularly important when dealing with building materials in section as described later in the chapter.

Equally important is the necessity to indicate clearly where the section is taken. This is done using a section line as shown in Fig. 5.8. As well as showing the location of the section, the section line indicates the direction of view. In practice the section line offers the draughtsman a degree of artistic licence. However, BS 1192 recommends the use of the section line shown. Finally, the draughtsman must not overlook the importance of relating the sectional view to the section line, viz. section A–A. This is of the utmost importance where many sectional views are necessary in such work as reinforced concrete detailing, where nearly all views are based on cross sections.

Figure 5.9 shows the same object by a horizontal cutting plane. The arrangement of views is shown in Fig. 5.9b.

Building sections

With the diversified and numerous materials used in the construction industry, it is particularly useful to be able to illustrate these materials in sectional views, since it is impractical to show them on elevational views. Standard methods of representation for building materials are shown in Fig. 5.10. These symbols are used **only where a material is shown in section**. The symbols shown are based upon the recommendations of BS 1192 and PD 6479.

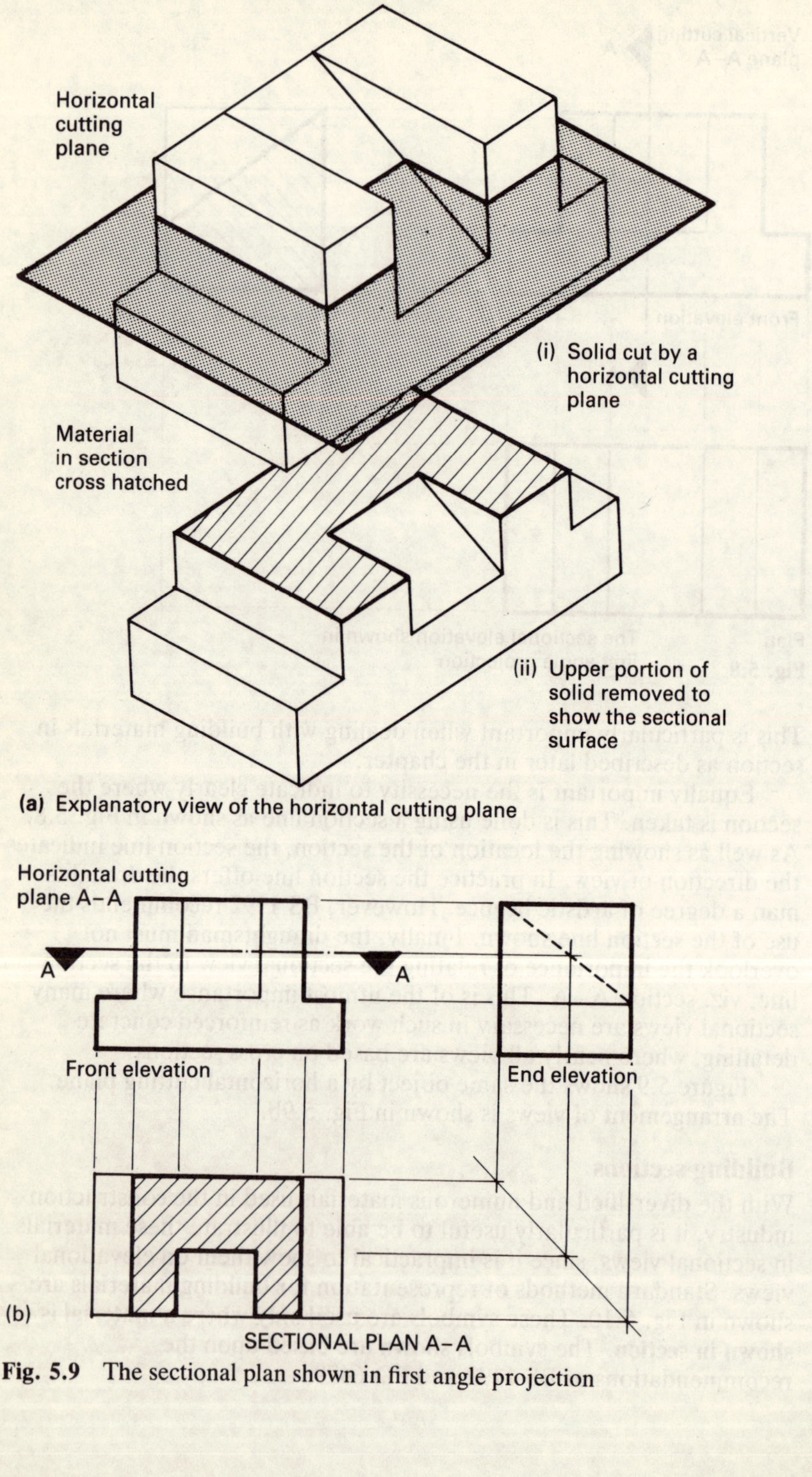

Fig. 5.9 The sectional plan shown in first angle projection

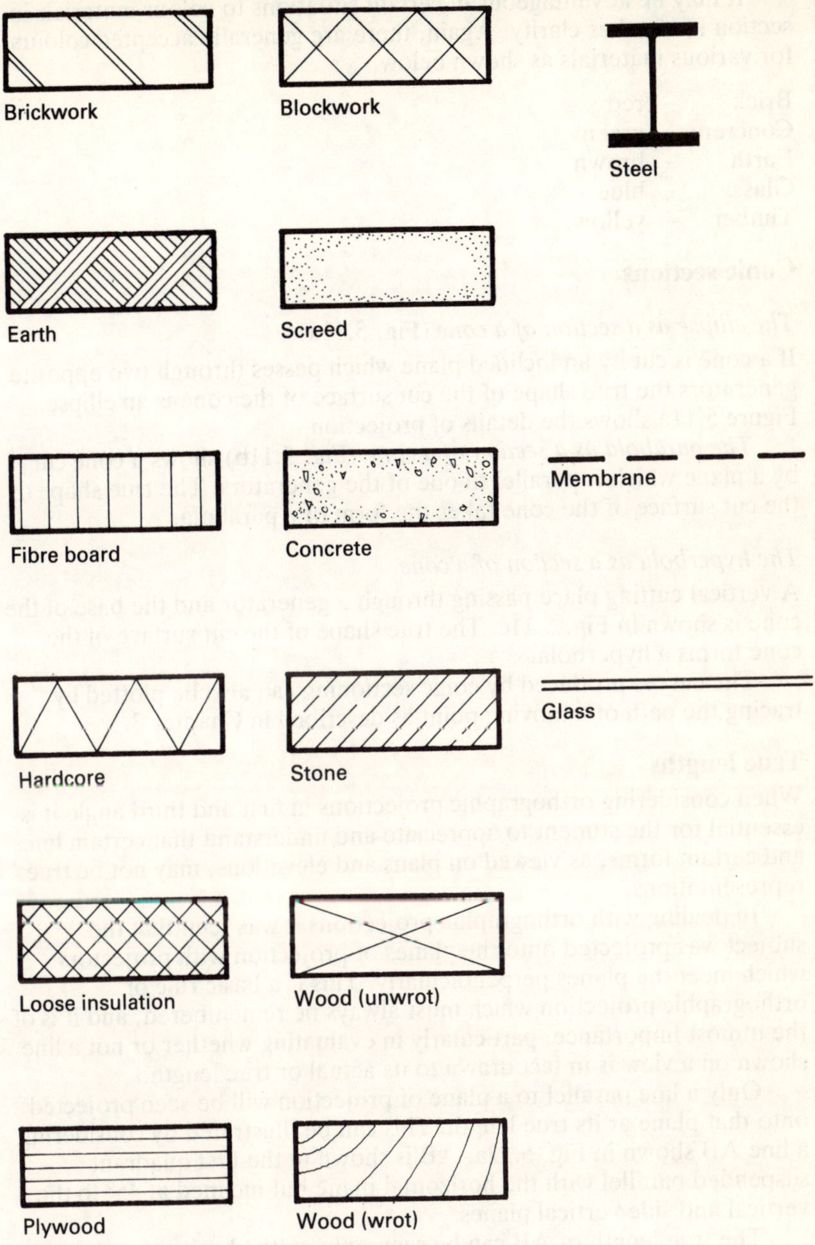

Fig. 5.10 Building materials in section

It may be advantageous in certain situations to colour materials in section for further clarity. Again, there are generally accepted colours for various materials as shown below:

Brick – red
Concrete – green
Earth – brown
Glass – blue
Timber – yellow.

Conic sections

The ellipse as a section of a cone (Fig. 5.11a)
If a cone is cut by an inclined plane which passes through two opposite generators the true shape of the cut surface of the cone is an ellipse. Figure 5.11a shows the details of projection.

The parabola as a section of a cone (Fig. 5.11b) shows a cone cut by a plane which is parallel to one of the generators. The true shape of the cut surface of the cone takes the form of a parabola.

The hyperbola as a section of a cone
A vertical cutting place passing through a generator and the base of the cone is shown in Fig. 5.11c. The true shape of the cut surface of the cone forms a hyperbola.

The curves produced by conic sectioning can also be plotted by tracing the path of a moving point as described in Chapter 3.

True lengths

When considering orthographic projections in first and third angle it is essential for the student to appreciate and understand that certain lines and certain forms, as viewed on plans and elevations, may not be true representations.

In dealing with orthographic projections it was seen that the subject was projected onto the planes of projection with projectors which meet the planes perpendicularly. This is a basic rule of orthographic projection which must always be remembered, and it is of the utmost importance, particularly in evaluating whether or not a line shown on a view is in fact drawn to its actual or true length.

Only a line parallel to a plane of projection will be seen projected onto that plane at its true length. This can be illustrated by considering a line AB shown in Fig. 5.12a. AB is shown in the first quadrant suspended parallel with the horizontal plane but inclined at 45° to the vertical and side vertical planes.

The true length of AB can be seen only on the horizontal plane since this is the only plane parallel with the line.

The V.P. and the S.V.P. show the length AB foreshortened due to the inclination of the line to these planes. Note that the projectors still meet the plane perpendicularly, thus projecting the foreshortened line

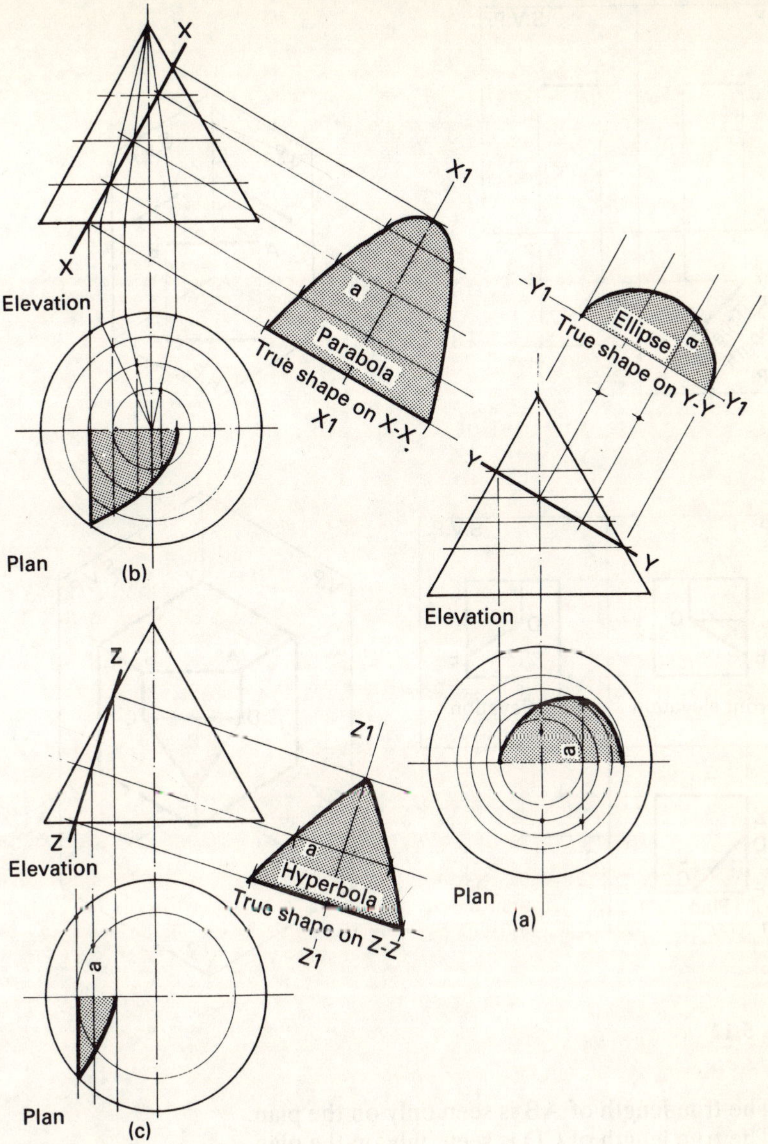

Fig. 5.11

length. In fact, the case illustrated shows the line length as $\frac{AB}{\sqrt{2}}$ since the line makes an angle of 45° with the planes.

Extending this principle to Fig. 5.12b, which shows a partly cutaway cube, it will be seen that:

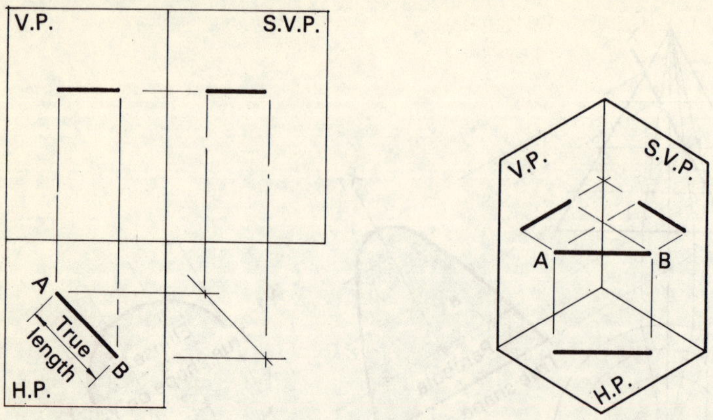

(a)

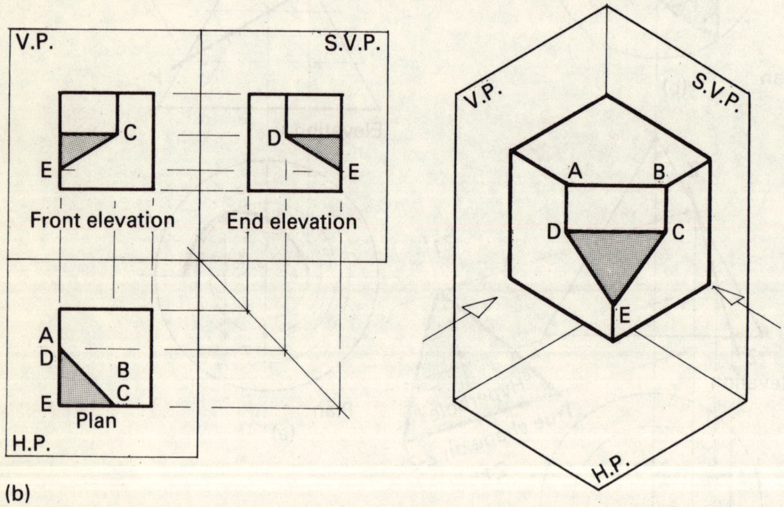

(b)

Fig. 5.12

- The true length of AB is seen only on the plan.
- The true length of CD is seen only on the plan.
- The true length of EC is seen only on the front elevation.
- The true length of ED is seen only on the end elevation.

In each case the line in question is parallel only with its respective plane.

Consider now the triangle CDE. In no case is the triangular plane CDE shown in its true form, since its surface is not parallel with any of the planes in question. In this case, the true shape of CDE can be determined by employing an auxiliary view as dealt with later in the chapter, or alternatively it can be plotted since the true lengths of the three enclosing sides can be determined from the projected views.

Auxiliary projection

To enable a clearer understanding of the shape or detail of an object it is often useful to employ planes of projection other than the principal planes. There are in fact an infinite number of alternative planes which may be utilized by the draughtsman, and these planes are known as auxiliary planes of projection, being auxiliary to the principal planes already described earlier in the chapter.

The object in using an auxiliary plane is to allow the draughtsman the facility of showing a true length, a true shape, or simply to produce a view from a specific angle.

To illustrate the principle of the auxiliary plane consider again the object in Fig. 5.13a suspended in the first quadrant, with face A parallel with the side vertical plane, face C parallel with the vertical plane and face D parallel with the horizontal plane. It should now be appreciated by the student that none of the principal views, when drawn, will indicate the true *shape* of the surface B, since none of the principal planes are parallel to it. If it is required to show the true shape of surface B an auxiliary plane can be established parallel to it such that when the surface is projected onto it the true shape will be seen. The plane shown in Fig. 5.13a is known as an auxiliary vertical plane, in this case being perpendicular to the horizontal plane and making the desired angle with the vertical plane.

The planes are now opened out into the vertical plane about XY and X′Y′ to produce an arrangement of planes as shown in Fig. 5.13b.

It can be seen from Fig. 5.13b that by projecting at 90° to face B the view is produced on the auxiliary vertical plane. This view is known as the *Auxiliary elevation*.

Note that the length of the auxiliary elevation is projected from the plan and heights taken from the elevation above XY.

For practical drawing purposes, to produce an auxiliary elevation an X′Y′ line should be produced parallel to the feature to be projected. The length of the feature is projected from the plan, and the height plotted from the elevation, bearing in mind that the distance of the object above X′Y′ is the same as its distance above XY. This means that if X′Y′ and XY are considered as datum lines, heights such as a and b will be equal above each datum. This is particularly worthwhile remembering when an auxiliary elevation of a more complex object is required.

An auxiliary plan can be drawn in much the same way by constructing a further auxiliary plane as shown in Fig. 5.14a. This plane is known as the auxiliary inclined plane.

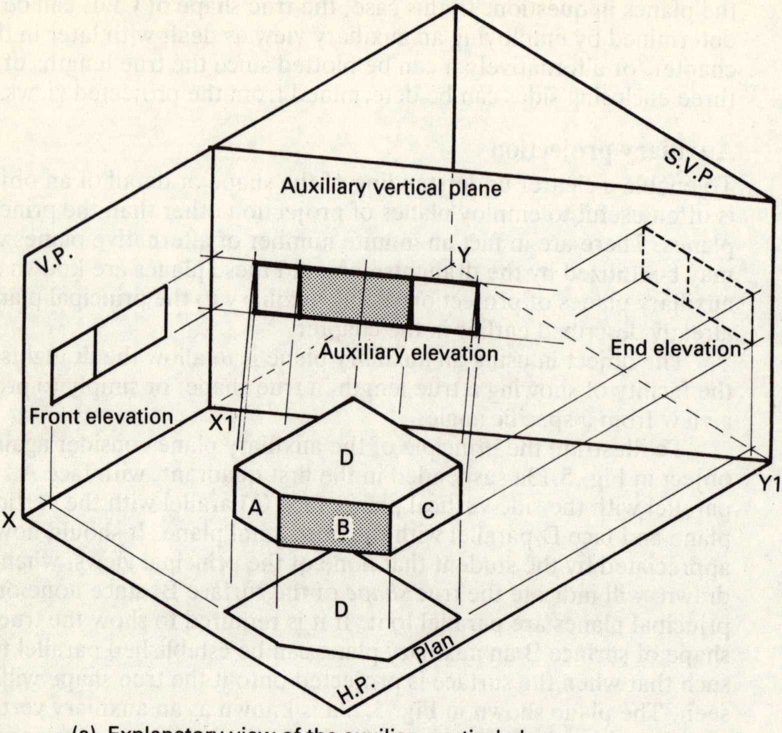

(a) Explanatory view of the auxiliary vertical plane

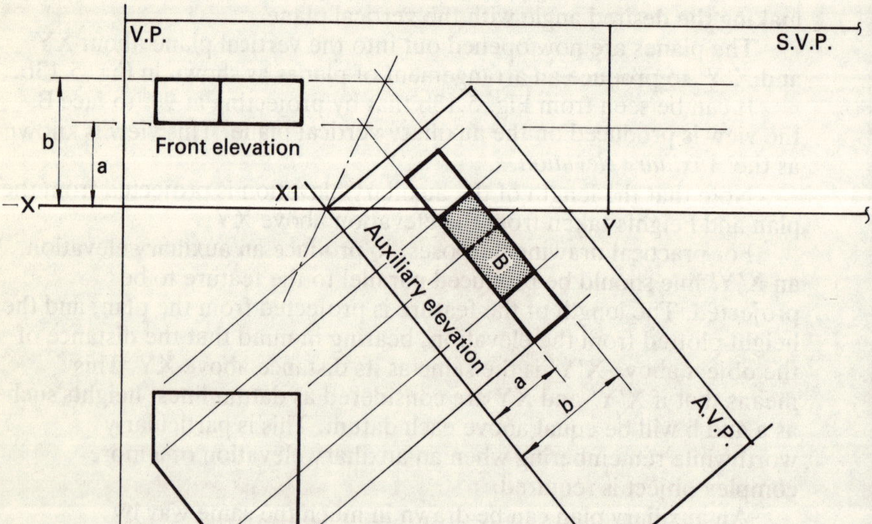

Fig. 5.13 (b) Arrangement of views in first angle projection

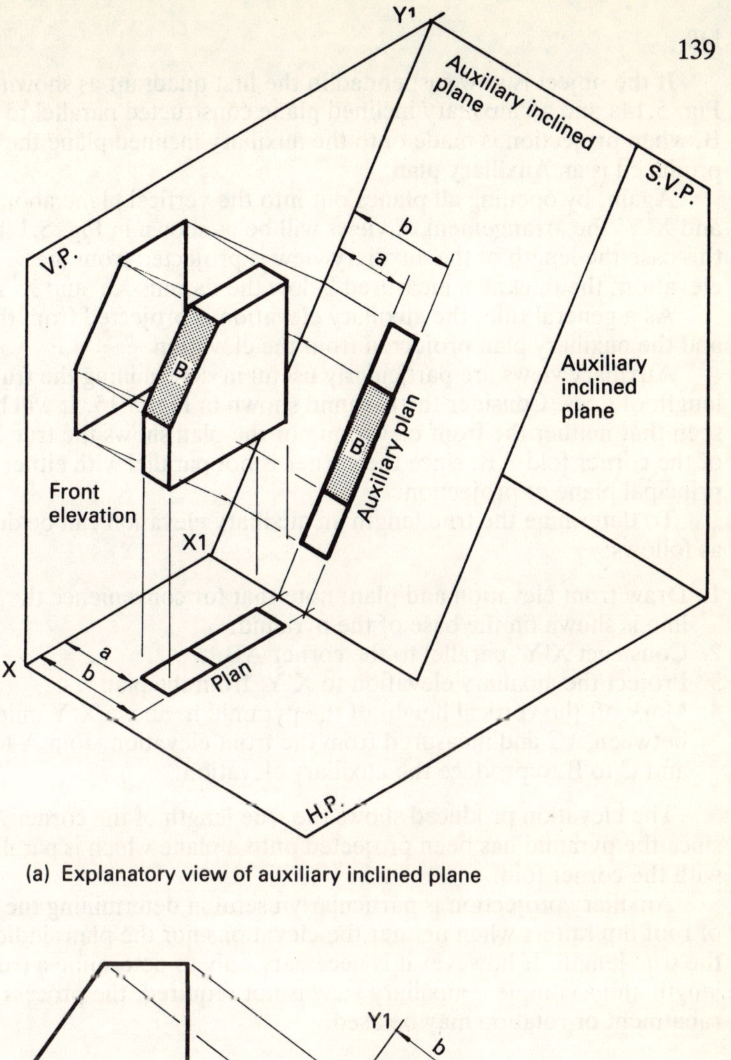

(a) Explanatory view of auxiliary inclined plane

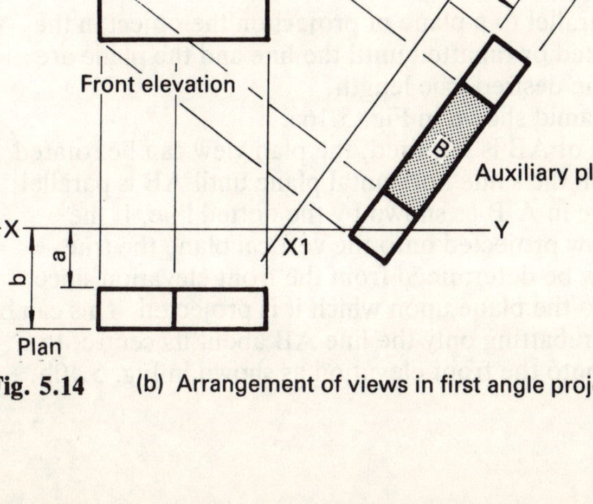

Fig. 5.14 (b) Arrangement of views in first angle projection

If the object is now suspended in the first quadrant as shown in Fig. 5.14a and an auxiliary inclined plane constructed parallel to face B, when projection is made onto the auxiliary inclined plane the view produced is an **Auxiliary plan**.

Again, by opening all planes out into the vertical plane about XY and X′Y′ the arrangement of views will be as shown in Fig. 5.14b. In this case the length of the auxiliary view is projected from the elevation; the thickness measured below the datums XY and X′Y′.

As a general rule, the auxiliary elevation is projected from the plan and the auxiliary plan projected from the elevation.

Auxiliary views are particularly useful in determining the true length of lines. Consider the pyramid shown in Fig. 5.15. It will be seen that neither the front elevation nor the plan shows the true length of the corner fold AB, since the corner is not parallel with either principal plane of projection.

To determine the true length an auxiliary elevation can be drawn as follows:

1. Draw front elevation and plan; note that for convenience the XY line is shown on the base of the pyramid.
2. Construct X′Y′ parallel to the corner AB.
3. Project the auxiliary elevation to X′Y′ from the plan.
4. Mark off the vertical height of the pyramid h, above X′Y′ midway between AC and measured from the front elevation. Join A to B and C to B to produce the auxiliary elevation.

The elevation produced shows the true length of the corner AB, since the pyramid has been projected onto a plane which is parallel with the corner fold.

Auxiliary projection is particularly useful in determining the length of roof hip rafters when neither the elevations nor the plans indicate the true length. If however it is necessary only to determine a true length and a complete auxiliary view is not required, the process of rabatment or rotation may be used.

Rabatment

When a line is not parallel to a plane of projection the object in the quadrant can be rotated or rabatted until the line and the plane are parallel to produce the desired true length.

Consider the pyramid shown in Fig. 5.16a.

If the true length of AB is required, the plan view can be rotated about its centre line in the same horizontal plane until AB is parallel with the vertical plane in A′B as shown by the dotted line. If the rotated pyramid is now projected onto the vertical plane the true length of AB can now be determined from the front elevation since A′B is now parallel to the plane upon which it is projected. This can be further simplified by rabatting only the line AB about its centre, to C, and then projecting onto the front elevation as shown in Fig. 5.10b.

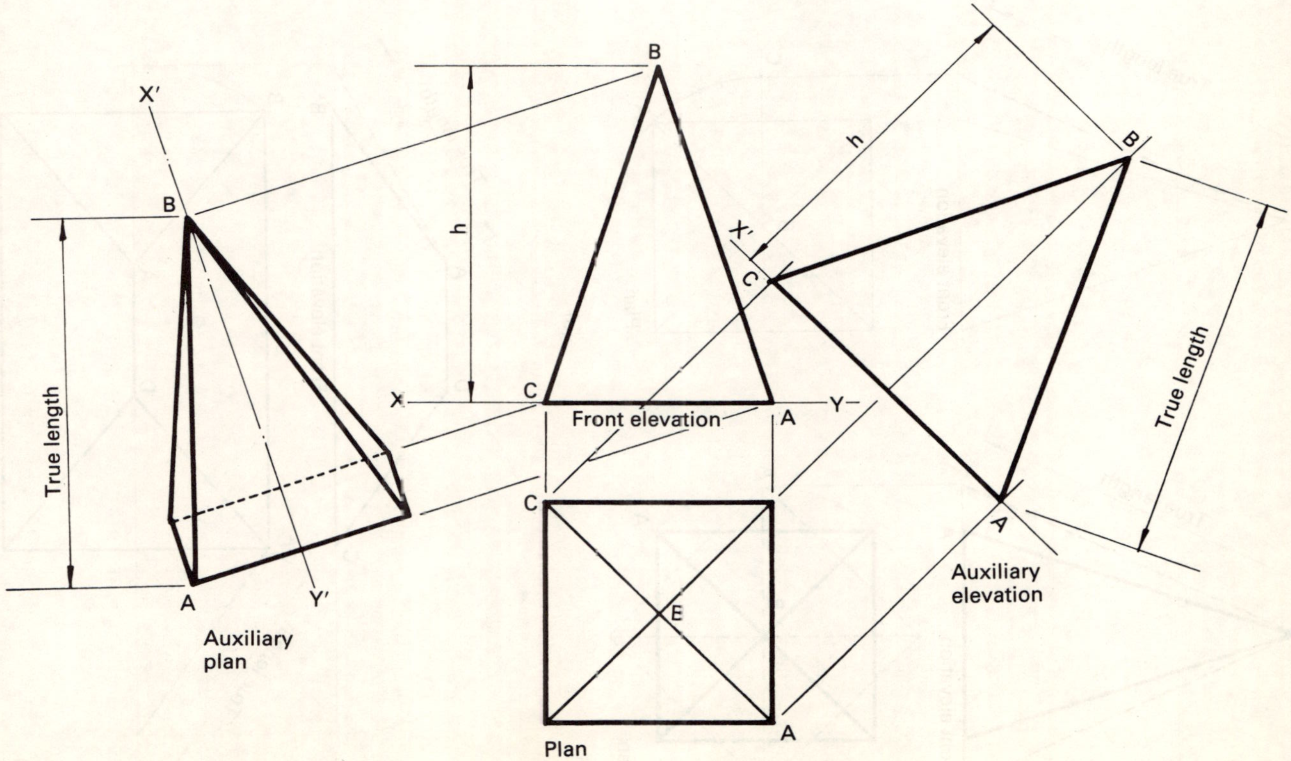

Fig. 5.15 Auxiliary projection

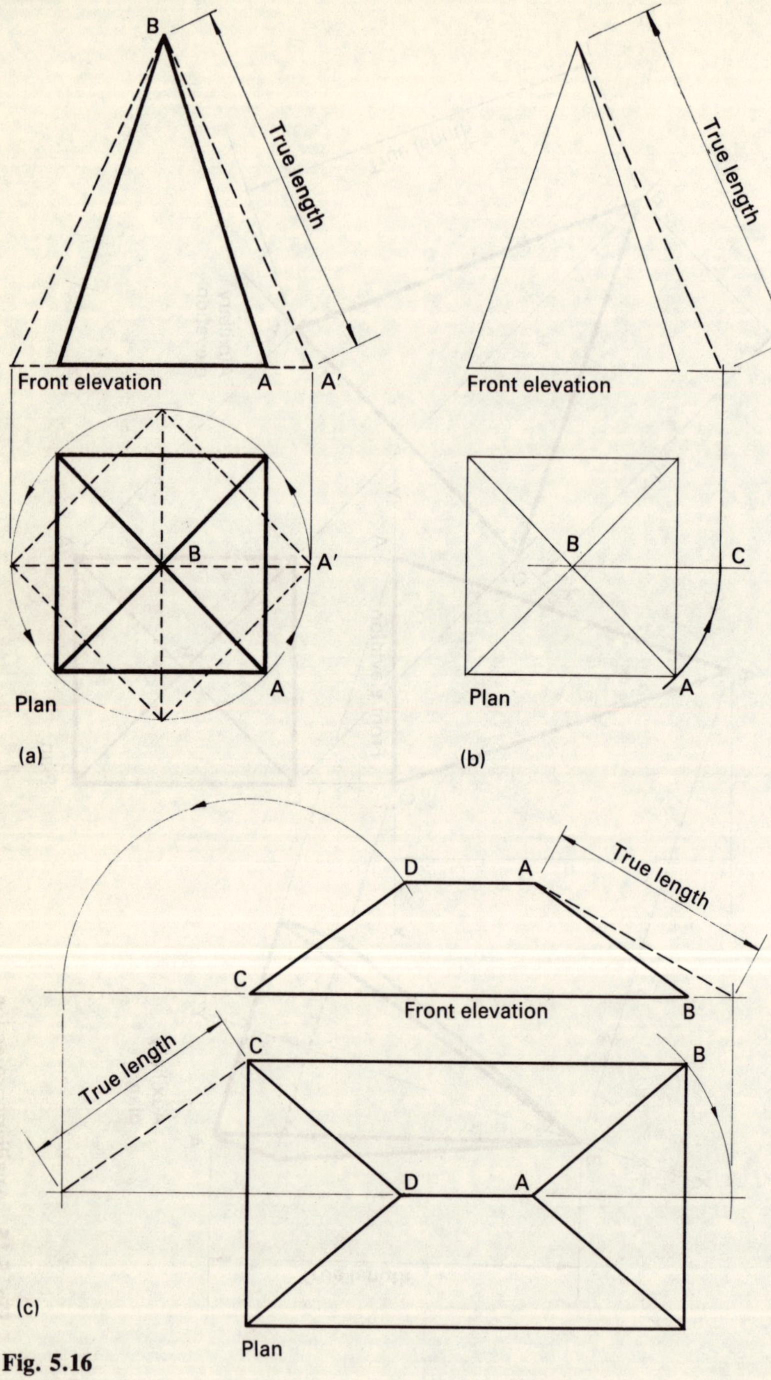
B
True length
Front elevation
A
A′
B
A′
Plan
A
(a)
True length
Front elevation
B
C
Plan
A
(b)
D
A
True length
C
Front elevation
B
True length
C
B
D
A
(c)
Plan

Fig. 5.16

The application of rabatment to roof geometry is shown in Fig. 5.16c. It can be seen that the hip AB can be either rotated on the plan and projected onto the elevation, or the true length of rafter CD seen on the elevation can be rotated on the elevation and projected onto the plan.

Exercises

1. Figure 5.17a shows a cylinder cut by a plane at 45°. In first angle projection draw the given front elevation, the plan and the end elevation viewed from arrow A. Scale full size.

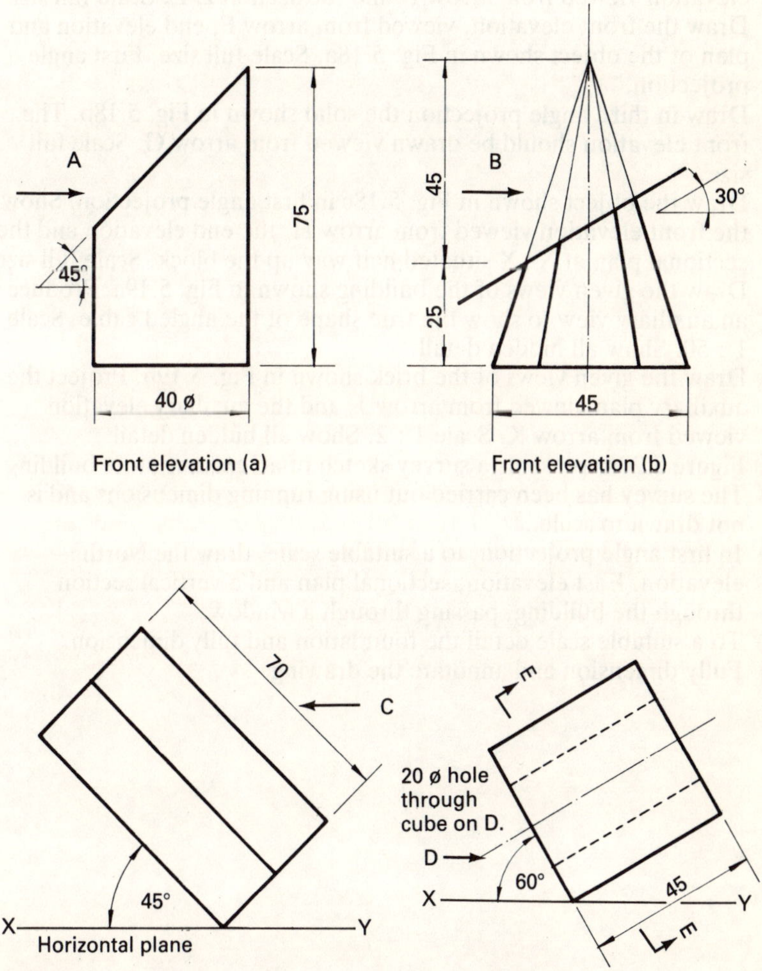

Fig. 5.17 Front elevation (c) Plan (d)

2. The truncated hexagonal pyramid shown in Fig. 5.17b is cut by a plane inclined at 30° to the horizontal. Draw in third angle projection the given front elevation, the plan, the end elevation viewed from arrow B, and an auxiliary view to show the true shape of the cut surface. Scale full size.
3. The regular hexagonal prism shown in Fig. 5.17c rests in the first quadrant, and makes an angle of 45° with the horizontal plane. Draw in first angle projection the given front elevation, the plan and the end elevation viewed from arrow C. Scale full size.
4. Figure 5.17d shows the plan view of a cube situated in the third quadrant, and making an angle of 30° with the vertical plane. Draw in third angle projection the given plan, the front elevation, the end elevation viewed from arrow D and the section E-E. Scale full size.
5. Draw the front elevation, viewed from arrow F, end elevation and plan of the object shown in Fig. 5.18a. Scale full size. First angle projection.
6. Draw in third angle projection the solid shown in Fig. 5.18b. The front elevation should be drawn viewed from arrow G. Scale full size.
7. Draw the object shown in Fig. 5.18c in first angle projection. Show the front elevation viewed from arrow H, the end elevation and the sectional plan at X–X situated half way up the block. Scale full size.
8. Draw the given views of the building shown in Fig. 5.19a. Produce an auxiliary view to show the true shape of the angled gable. Scale 1 : 50. Show all hidden detail.
9. Draw the given views of the brick shown in Fig. 5.19b. Project the auxiliary plan viewed from arrow J, and the auxiliary elevation viewed from arrow K. Scale 1 : 2. Show all hidden detail.
10. Figure 5.20 represents a survey sketch of an existing store building. The survey has been carried out using running dimensions and is not drawn to scale.
    (a) In first angle projection, to a suitable scale, draw the North elevation, East elevation, sectional plan and a vertical section through the building, passing through a window.
    (b) To a suitable scale detail the foundation and fully dimension.
    (c) Fully dimension and annotate the drawing.

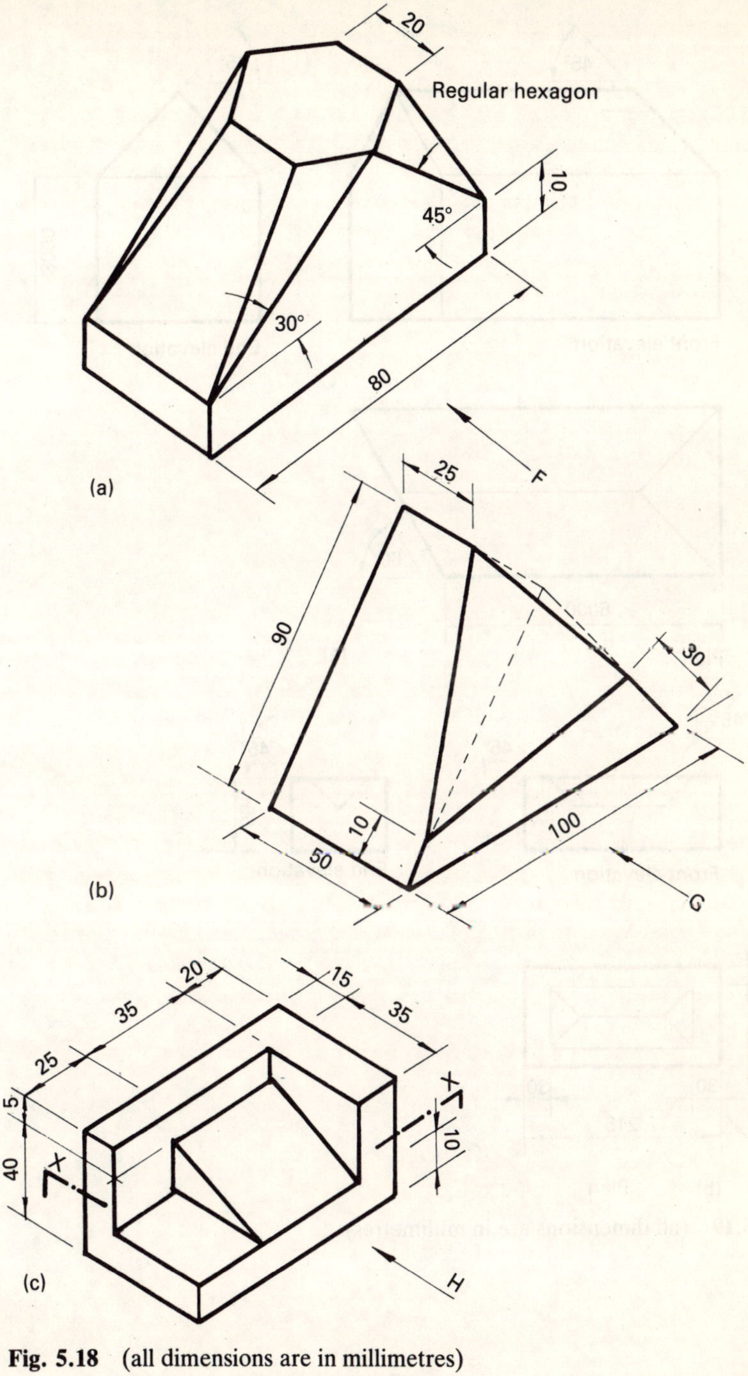

Fig. 5.18 (all dimensions are in millimetres)

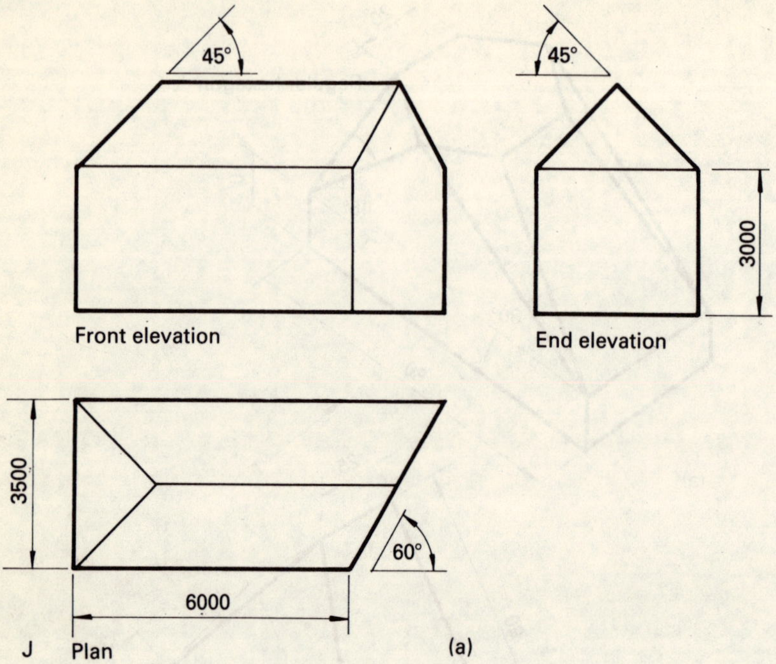

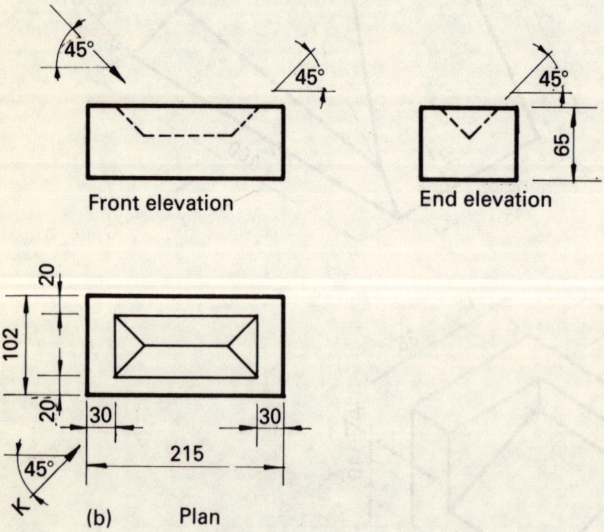

Fig. 5.19 (all dimensions are in millimetres)

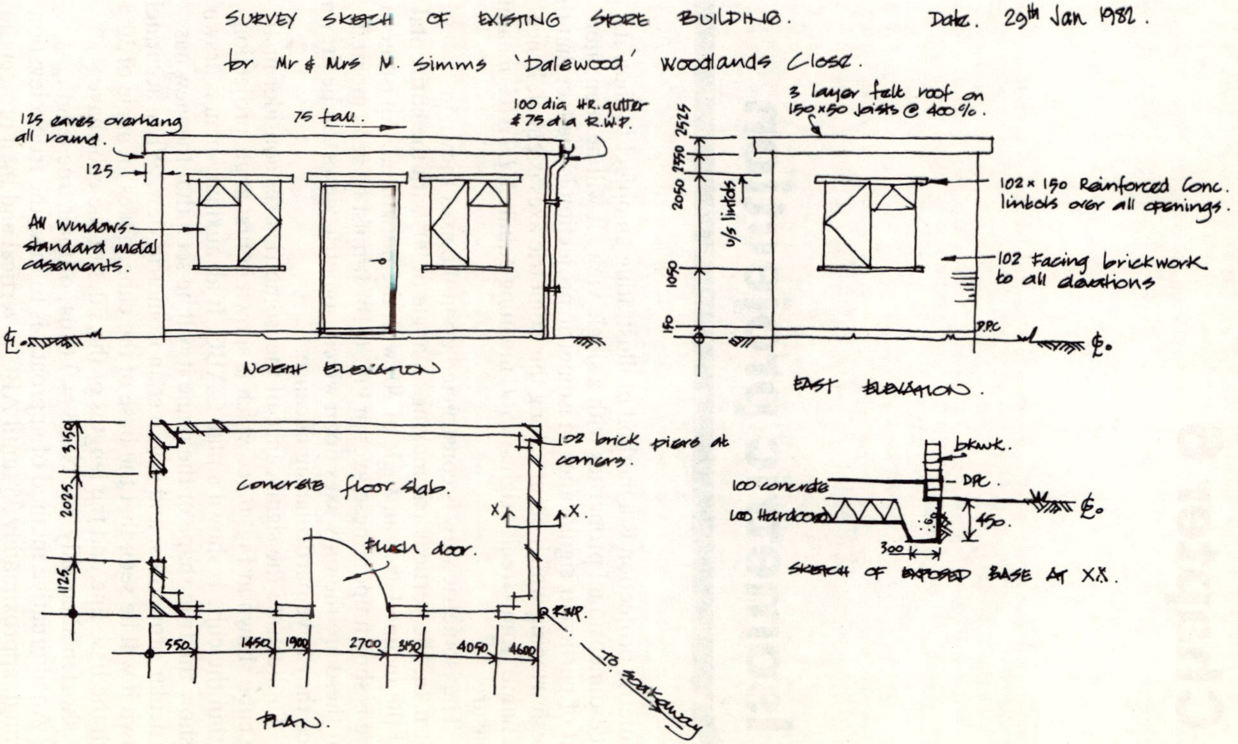
SURVEY SKETCH OF EXISTING STORE BUILDING.
Date. 29th Jan 1982.
for Mr & Mrs N. Simms 'Dalewood' Woodlands Close.
125 eaves overhang all round.
75 fall.
100 dia HR. gutter & 75 dia R.W.P.
125
All windows standard metal casements.
NORTH ELEVATION
3 layer felt roof on 150×50 joists @ 400 %.
102 × 150 Reinforced Conc. lintols over all openings.
102 Facing brickwork to all elevations
2525
2350
2050
u/s lintols
1050
150
D.P.C
EAST ELEVATION.
102 brick piers at corners.
concrete floor slab.
flush door.
3150
2025
1125
550
1450
1900
2700
3150
4050
4600
R.W.P.
TO SOAKAWAY
PLAN.
bkwk.
DPC.
100 concrete
100 Hardcore
450.
300
SKETCH OF EXPOSED BASE AT X.X.
Fig. 5.20

Chapter 6

Isometric projection

Isometric projection is a method of illustrating an object to show its three-dimensional properties with a single view. It will be seen later that of particular significance in isometric projection is the fact that the foreshortening which occurs along the isometric axes is equal, hence the name from Greek origins; *ISO*, meaning equal; *METRIC*, meaning measure.

The isometric view is constructed upon the axes shown in Fig. 6.1a, the vertical axis making an angle of 90° with the horizontal and the inclined axes an angle of 30° with the horizontal. An isometric cube is shown upon these axes in Fig. 6.1b. It must be remembered that these are the only axes upon which isometric views can be drawn, since they have particular significance.

To illustrate the significance of the isometric axes consider firstly the cube shown in Fig. 6.1c, which is shown in first angle projection. Within the cube is shown a plane, ABC. If an auxiliary plan is drawn to show the true shape of the plane it will be seen that the view has particular properties, and is in fact an isometric view. When accurately drawn it will be seen that the base of the cube makes an angle of 30° with the base line and the lengths of the sides of the cube have foreshortened equally, i.e. $\sqrt{2/3} = 0.8164$, of their true length.

An alternative method of approach is to raise the front elevation through approximately 35° until ABC is vertical and then project the end elevation. The final result will be the same. This then means that an object inclined at this angle to the horizontal plane can be seen in isometric projection.

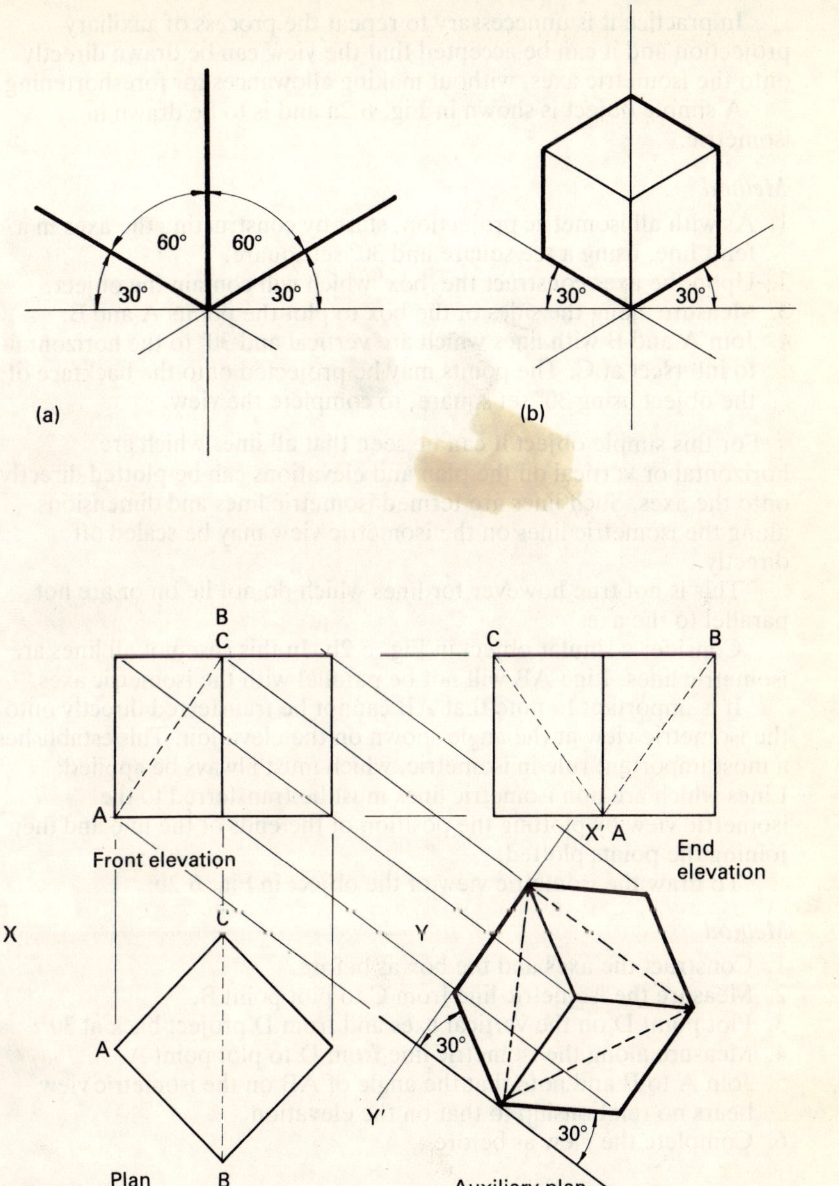

Fig. 6.1 Isometric projection

In practice it is unnecessary to repeat the process of auxiliary projection and it can be accepted that the view can be drawn directly onto the isometric axes, without making allowances for foreshortening.

A simple object is shown in Fig. 6.2a and is to be drawn in isometric.

Method

1. As with all isometric projection, start by constructing the axes in a feint line, using a tee square and 30° set square.
2. Upon the axes construct the 'box' which will contain the object.
3. Measure along the sides of the box to plot the points A and B.
4. Join A and B with lines which are vertical and 30° to the horizontal. to intersect at C. The points may be projected onto the backface of the object using 30° set square, to complete the view.

For this simple object it can be seen that all lines which are horizontal or vertical on the plan and elevations can be plotted directly onto the axes. Such lines are termed isometric lines and dimensions along the isometric lines on the isometric view may be scaled off directly.

This is not true however for lines which do not lie on or are not parallel to the axes.

Consider a similar object in Fig. 6.2b. In this case not all lines are isometric lines. Line AB will not be parallel with the isometric axes.

It is important to note that AB cannot be transferred directly onto the isometric view at the angle shown on the elevation. This establishes a most important rule in isometric, which must always be applied: Lines which are non isometric lines must be transferred to the isometric view by plotting the position of the ends of the line and then joining the points plotted.

To draw the isometric view of the object in Fig. 6.2b:

Method

1. Construct the axes and the box as before.
2. Measure the isometric line from C to plot point B.
3. Plot point D on the vertical axes and from D project back at 30°.
4. Measure along the isometric line from D to plot point A.
5. Join A to B and note that the angle of AB on the isometric view bears no relationship to that on the elevation.
6. Complete the view as before.

Circles

Circles drawn in isometric projection will appear on the isometric view as an ellipse having a major axis inclined at 60° to the horizontal.

To draw the isometric circle it is necessary to plot points on the circle circumference relative to a datum line and transfer these points onto the isometric view.

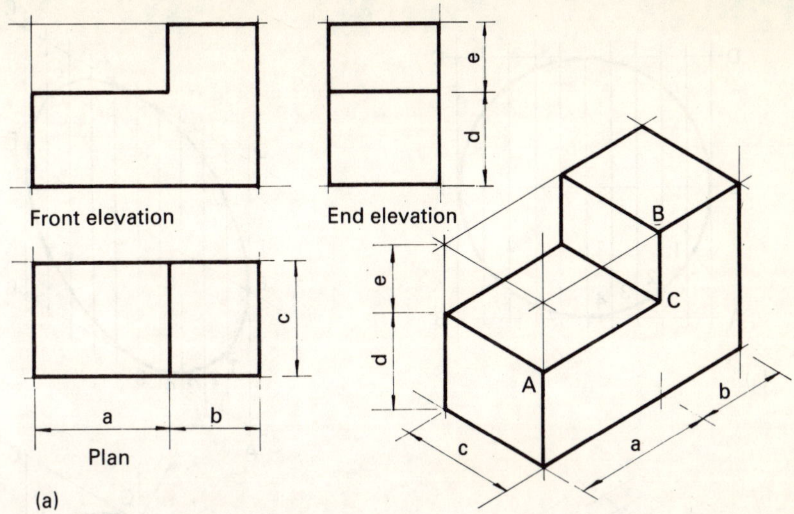

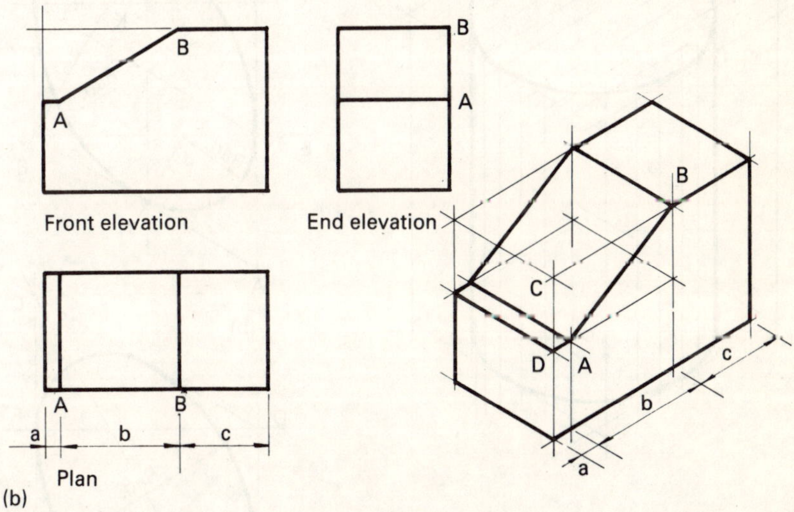

Fig. 6.2

To construct the isometric circle (Fig. 6.3a)

1. Draw the circle of given radius and enclose with a square ABCD of side length equal to the circle's diameter.
2. Divide the diameter of the circle EF into a convenient number of equal divisions – say 10 – and erect perpendiculars to intersect AB and CD at the points shown.

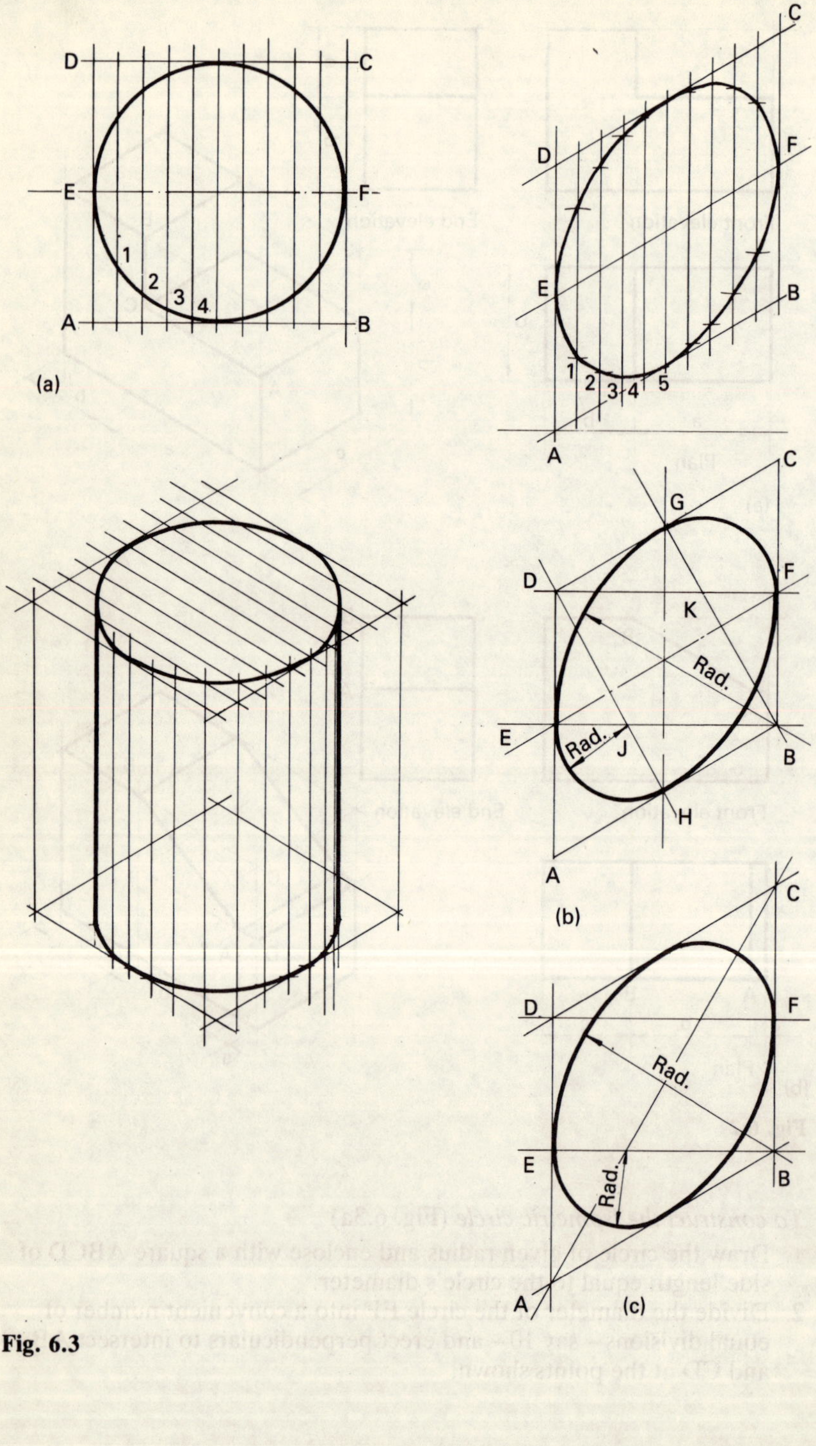
D
C
E
F
1
2
3
4
A
B
(a)
D
C
F
E
B
1
2
3
4
5
A
C
G
D
F
K
Rad.
E
Rad.
J
B
H
A
(b)
C
D
F
Rad.
E
B
Rad.
A
(c)

Fig. 6.3

3. Construct the isometric axes, and upon the axes construct the square ABCD and the centre lines of the circle.
4. Erect the perpendiculars along EF, dividing the square into 10 equal parts.
5. Using the centre line EF as a datum, mark off points 1, 2, 3, 4 and 5 in the first quadrant. The points may then be carefully joined, either freehand or using a curve, to form part of the isometric circle.

To rationalize the process, the points may be plotted in all quadrants simultaneously, since the relationship of the points to the datum line is constant in all four quadrants.

It also follows that in the initial process shown in 1 and 2 it is necessary to draw only one quadrant of the circle since all quadrants are symmetrical.

To further rationalize the process of drawing isometric circles, an approximation can be drawn using circular arcs, which obviates the necessity for freehand curves. These constructions are shown in Fig. 6.3b and c.

To construct an approximation of an isometric circle
Method 1 (Fig. 6.3b)

1. Draw the isometric square ABCD which will contain the circle, and the centre lines EF and GH.
2. Join F to D and E and G to H and B.
3. The points B and D can now be utilized as centres for the circular arcs of radius JE drawn about centres J and K, and radius BE drawn about centres B and D. The curves so drawn form an approximation to the isometric circle.

Method 2 (Fig. 6.3c)

1. Draw the isometric square ABCD to contain the circle.
2. Draw the diagonal AC.
3. Draw horizontal lines originating from B and D to intersect the diagonal at points G and H, giving the centres of the circular arcs as in method 1.

The method described for drawing the isometric circle can be applied to irregular shapes or arcs which have to be illustrated in isometric. As is often the case in isometric drawing, it will invariably be necessary to draw the plans and elevations before the isometric can be drawn.

Exercises Isometric projection

1. Figure 6.4 shows a sequence of solids, drawn in first angle projection. Construct the isometric view of each.

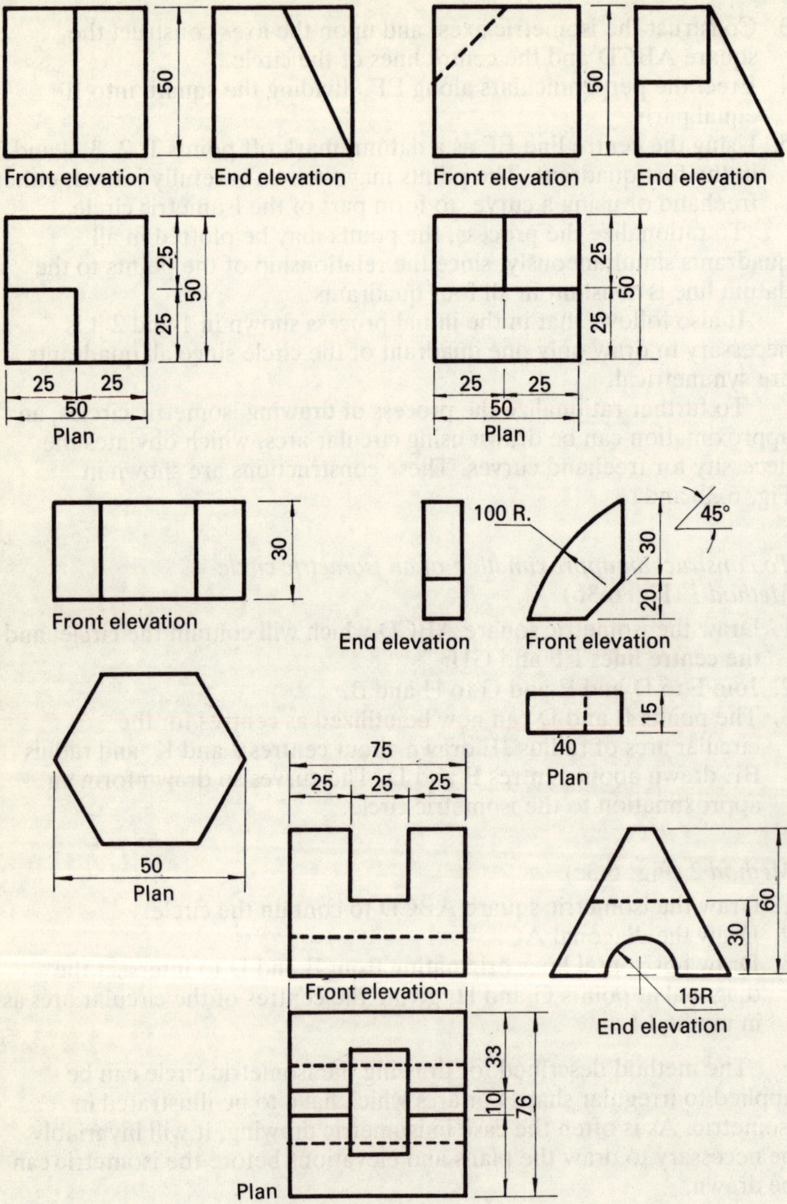

Fig. 6.4 Exercises

2. To a suitable scale, prepare an isometric view of a 1.5 m deep by 1 m wide foundation trench in moderate ground and timbered to the following specification: 225 × 38 poling boards; 150 × 100 walings; 100 × 100 struts. Wedges as required.
Illustrate a minimum trench length of 3 m.

Chapter 7

Oblique projection

Like isometric projection, oblique projection is based upon three axes shown in Fig. 7.1a. The axes in this case are horizontal, vertical and a receding axis which is generally at 45° to the horizontal, but may in certain cases be 30° or 60°.

The fundamental principle underlying oblique projection is that the projected view is based upon a true elevation contained within the horizontal and vertical axes, and all horizontal lines which do not form part of the true elevation are projected along or parallel with the receding axis. Figure 7.1b illustrates the principle by showing a cube of 50 mm side drawn in oblique. Note that the front face of the cube ABCD is a true elevation, as indeed is the back face, and the remaining faces are projected back at 45°.

It can be seen that the view appears elongated because of the lack of perspective (the apparent converging of lines at a distance). For this reason it is often preferable to use the longest elevation of an object as the true elevation and project the shorter sides along the receding axis as illustrated by Fig. 7.1c. It follows also that the drawing may be considerably simplified by taking the most complex view as the true elevation, since this may obviate the necessity to plot circles, curves, angles, etc. on the receding axis. This is often justified, in that oblique projection is generally employed where an elevation is particularly important.

Oblique projection may be considered under two headings.

(c) Cavalier

(d) Cabinet

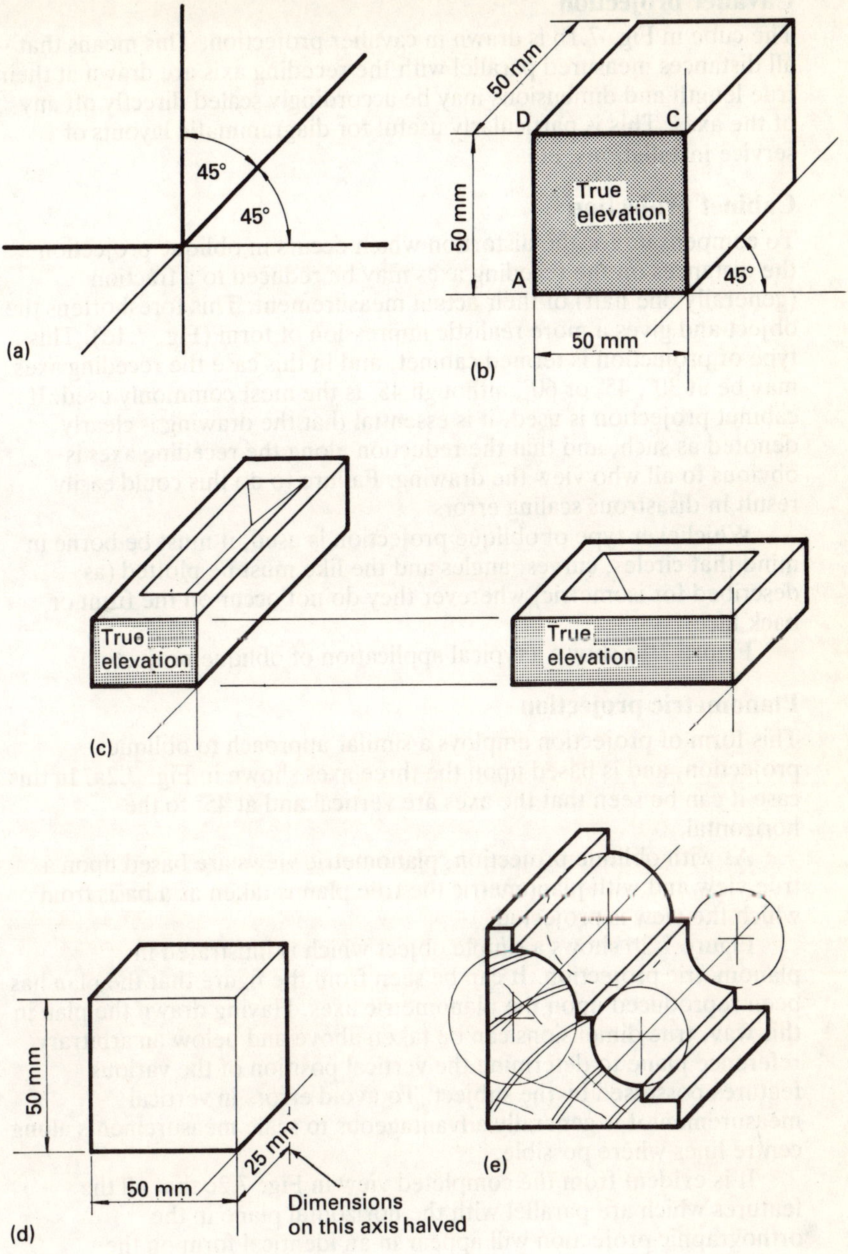

Fig. 7.1 Oblique projection

Cavalier projection

The cube in Fig. 7.1b is drawn in cavalier projection. This means that all distances measured parallel with the receding axis are drawn at their true length and dimensions may be accordingly scaled directly off any of the axes. This is particularly useful for diagrammatic layouts of service installations, etc.

Cabinet projection

To compensate for the distortion which occurs in oblique projection, the distances on the receding axes may be reduced to a fraction (generally one half) of their actual measurement. This foreshortens the object and gives a more realistic impression of form (Fig. 7.1d). This type of projection is termed cabinet, and in this case the receding axes may be at 30°, 45° or 60°, although 45° is the most commonly used. If cabinet projection is used, it is essential that the drawing is clearly denoted as such, and that the reduction along the receding axes is obvious to all who view the drawing. Failure to do this could easily result in disastrous scaling errors.

Whichever type of oblique projection is used, it must be borne in mind that circles, curves, angles and the like must be plotted (as described for isometric) wherever they do not occur on the front or back face.

Figure 7.1e shows a typical application of oblique projection.

Planometric projection

This form of projection employs a similar approach to oblique projection, and is based upon the three axes shown in Fig. 7.2a. In this case it can be seen that the axes are vertical and at 45° to the horizontal.

As with oblique projection, planometric views are based upon a true view and with planometric the true plan is taken as a basis from which the view is projected.

Figure 7.2b shows a simple object which is illustrated in planometric projection. It can be seen from the figure that the plan has been reproduced upon the planometric axes. Having drawn the plan in this way, true dimensions can be taken above and below an arbitrary reference plane to determine the vertical position of the various features possessed by the subject. To avoid errors in vertical measurement it is generally advantageous to scale measurements along centre lines where possible.

It is evident from the completed view in Fig. 7.2c that all the features which are parallel with the horizontal plane in the orthographic projection will appear in an identical form on the planometric view. This is illustrated clearly by the circles in the figure, which in every case are drawn as true circles. Lines which are not vertical or horizontal in orthographic projection must be positioned by plotting the extremities of the line and joining the points obtained.

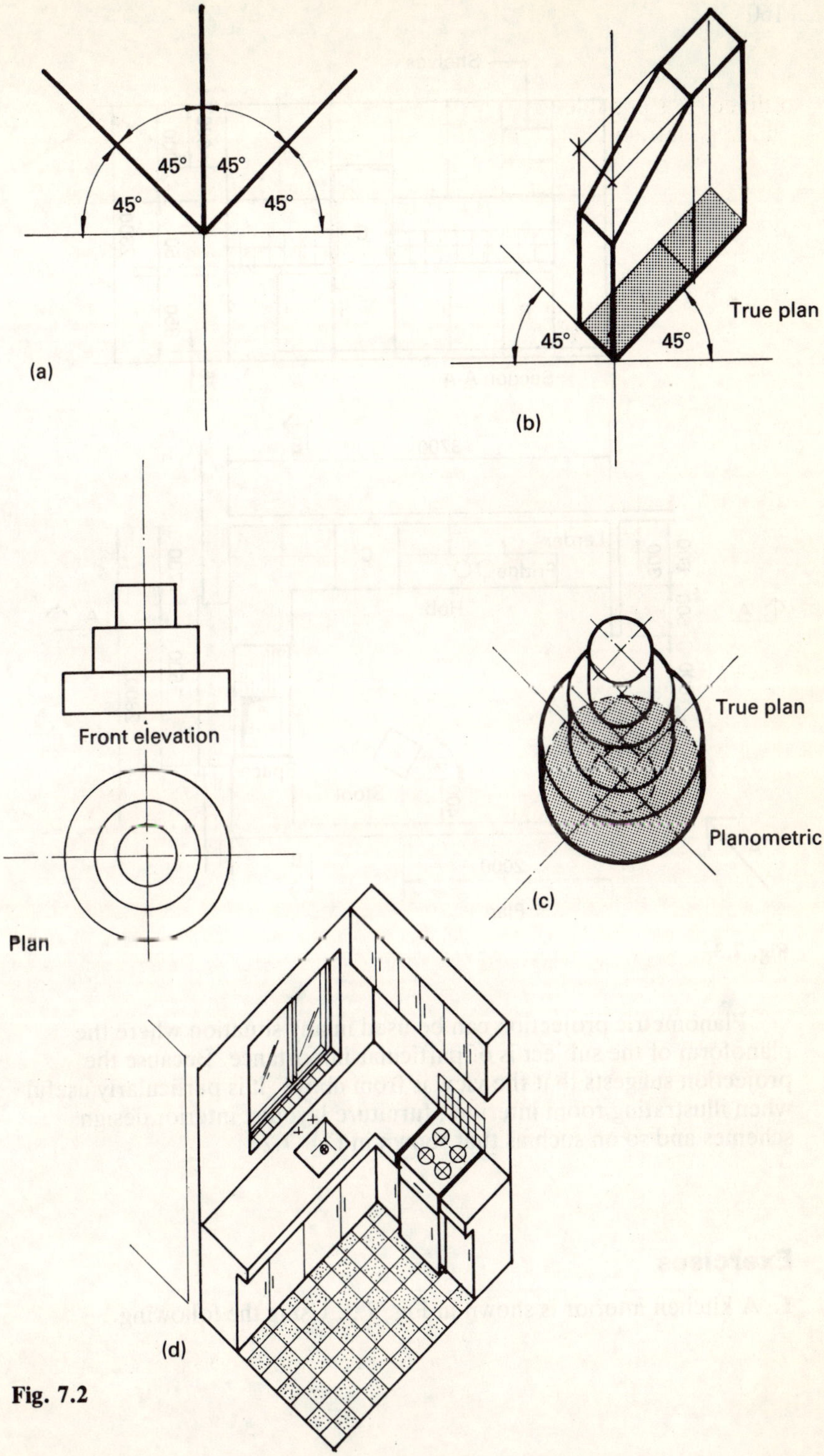
45°
45°
45°
45°
(a)
45°
45°
True plan
(b)
Front elevation
Plan
True plan
Planometric
(c)
(d)

Fig. 7.2

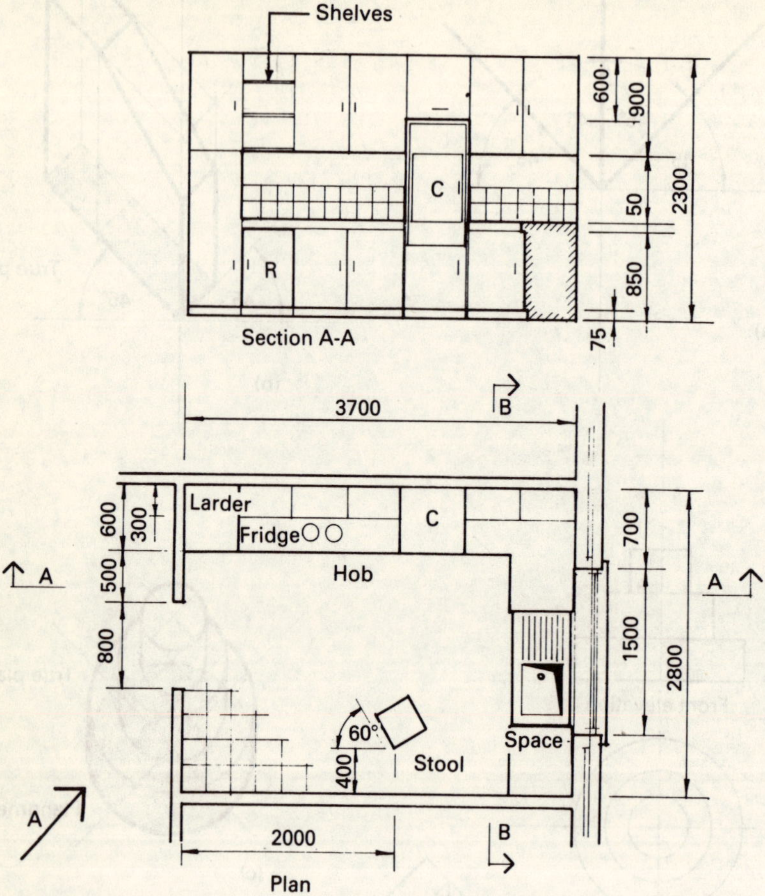

Fig. 7.3

Planometric projection can be used in any situation where the planoform of the subject is of particular importance. Because the projection suggests that the view is from above, it is particularly useful when illustrating room interiors, furniture layouts, interior design schemes and so on such as that shown in Fig. 7.2d.

Exercises

1. A kitchen interior is shown in Fig. 7.3. Using the following

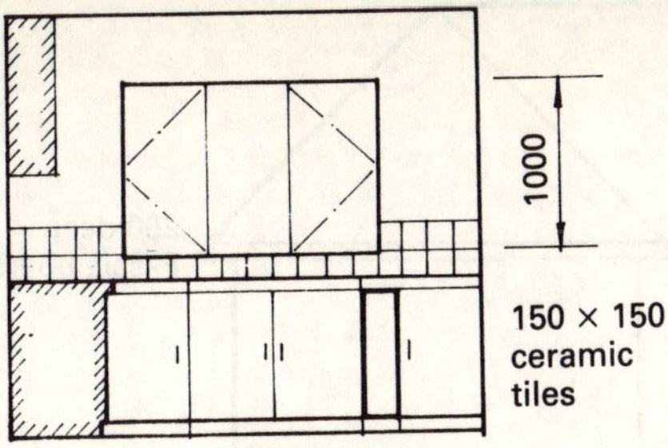

Section B-B

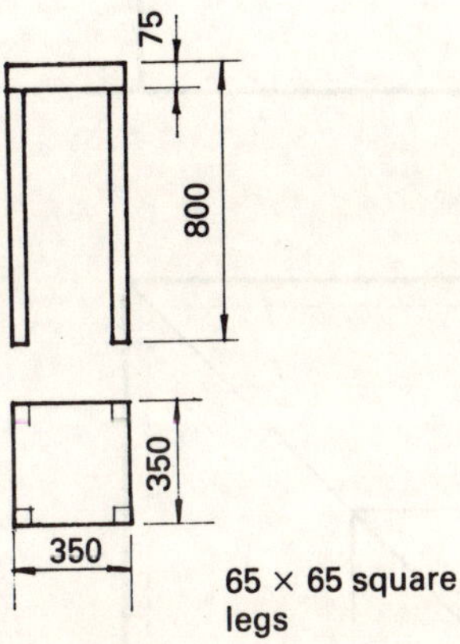

Stool detail

information together with the drawing, produce a planometric of the interior viewed from the direction of arrow A.
Base cupboards   500 mm wide × 575 mm deep
Refrigerator   600 mm wide × 575 mm deep
Oven housing   600 mm wide × 575 mm deep
Worktop   Continuous worktop 600 mm wide × 50 mm thick.
Floor covering   225 mm × 225 mm tiles.
Scale 1 : 20.

2. Figure 7.4 shows the outline of a two-storey detached house. To a scale of 1 : 100, draw a planometric view looking in the direction of arrow A.

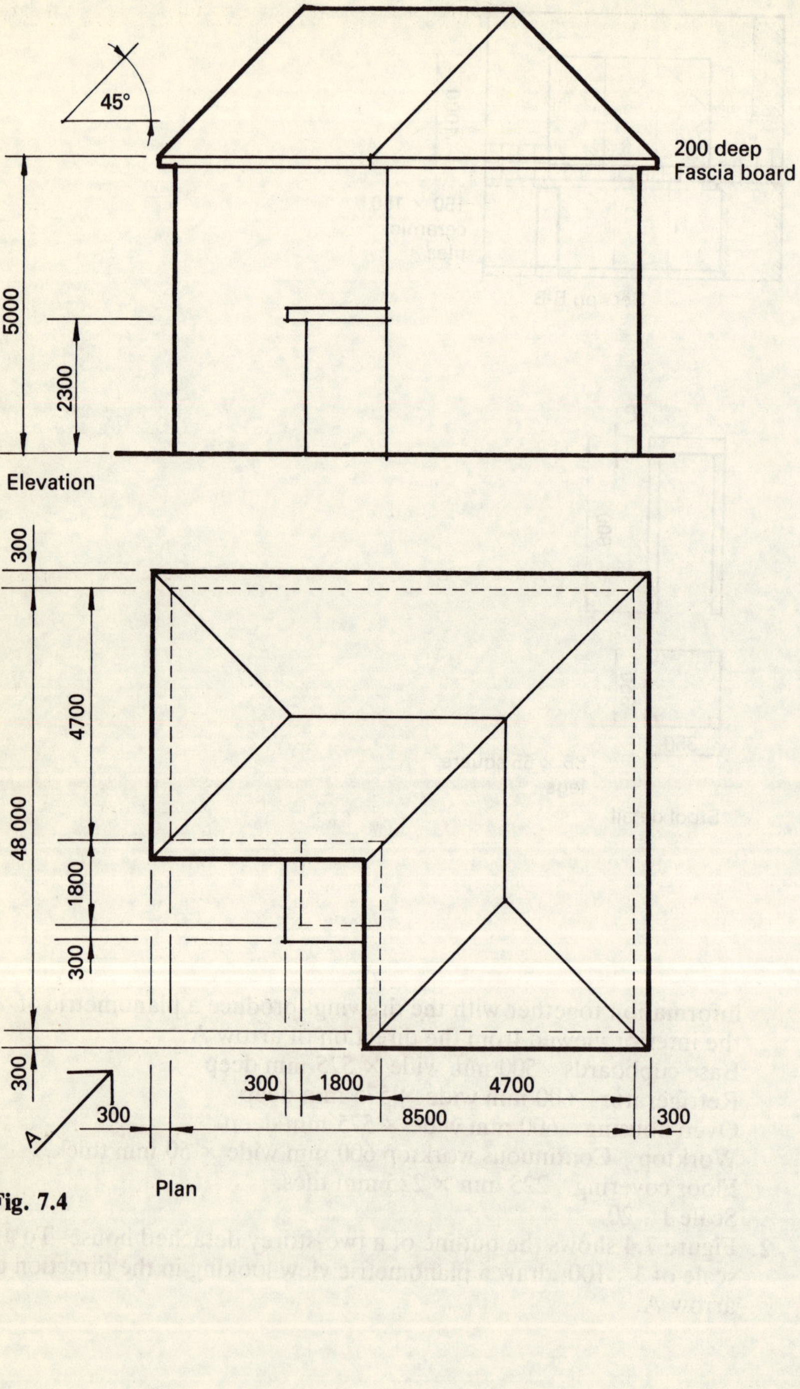
45°
200 deep
Fascia board
5000
2300
Elevation
300
4700
48 000
1800
300
300
300
1800
4700
300
8500
300
A
Plan

Fig. 7.4

Chapter 8

Developments

The term development describes the geometrical process of illustrating the surface of an object by laying out the surface on a single flat plane. Conversely, the developed surface, when folded along specific lines, will take the form of the three-dimensional object.

This can be illustrated by considering a simple form such as the cube. The cube has six identical square faces and twelve edges, all of equal length. If the cube is developed by opening out its surface, it will take the form of six squares, each square having a common fold line with its adjoining square (Fig. 8.1a). If the development were folded at right angles along each fold line, a cube would be produced.

Although the example given is quite basic, two important fundamentals of development have been established, since it can be seen that

(a) all lines on a development are true lengths; and consequently,
(b) all surfaces on a development are true shapes.

Furthermore, it can be seen that very often it is possible to gain a fairly accurate visual impression of a developed surface before actually commencing the drawing, and accordingly it may be useful to sketch roughly the foreseen shape, as a guide.

Further simple developments of prisms are shown in Fig. 8.1b–d and these require no further explanation.

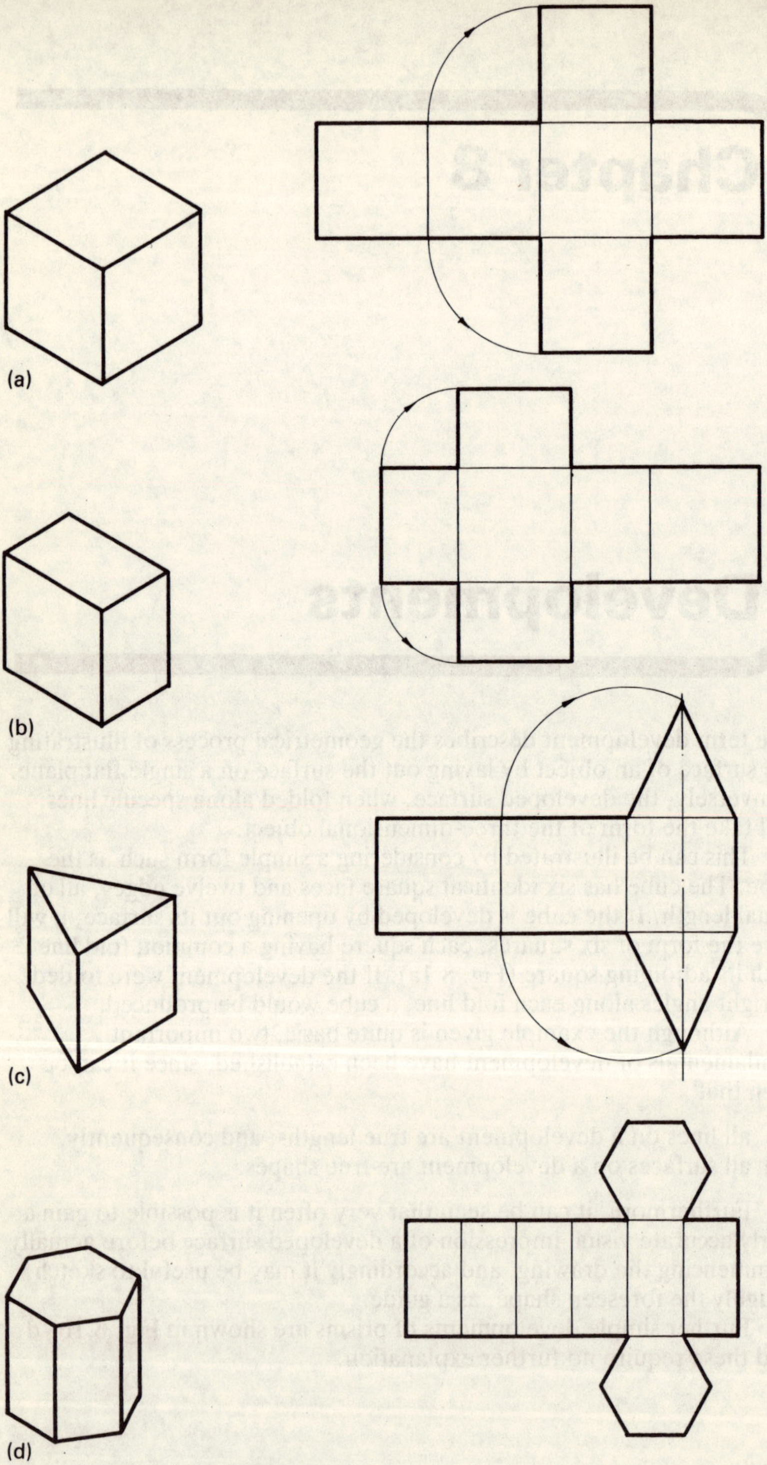

Fig. 8.1 Development of simple surfaces

Development of prisms

Following the simple principle already outlined, it is possible to consider the development of prisms which are penetrated by cutting planes.

Figure 8.2a shows a rectangular prism cut by a plane at 45° to the horizontal.

Method

1. Draw the elevation and plan of the prism and locate the cutting plane on the elevation.
2. Identify the relevant points a, b, c, d, e, f, g, and h at the base and top of the prism and 1, 2, 3, 4 at the cutting plane level.
3. Before commencing the development, the student should assure himself that all true lengths are known.
4. Project the developed surface of the sides of the prism, disregarding the cutting plane at this stage.
5. Index the developed surface a, b, c, d and e, f, g, h.
6. Project points 1, 2, 3, 4 from the elevation onto the development, ensuring that the points are projected onto the correct vertical, i.e. a1, b2, c3, d4.
7. Join points 1, 2, 3 and 4 to plot the line of the cutting plane.

If the top section of the prism is removed to produce a truncated prism, the top and bottom surfaces can be developed as shown. Note that the fold lines are shown in a thin continuous line.

This principle applies to the development of any truncated prism, and Fig. 8.2b shows the development of a truncated hexagonal prism.

The development of the cylinder (Fig. 8.3a)

Method

1. Draw plan and elevation of the cylinder.
2. Divide the plan into an equal number of divisions (12 is convenient), using a 60° × 30° set square.
3. Index the points 1 to 12 on the plan view.
4. The height of the developed surface can now be projected across from the elevation, and the length of the surface being equal to the circumference of the circle. This length can be determined geometrically by stepping the chord length 12–1 along the developed length 12 times using dividers. The top and bottom developments can be added, these being true circles.

It will be noted that in this case the chord distance is taken along the base of the development. This does result in small dimensional errors since the chord distance will be less than the arc distance (Fig. 8.3b). This is unlikely to be significant in student exercises, but if greater accuracy is required the developed length can be calculated.

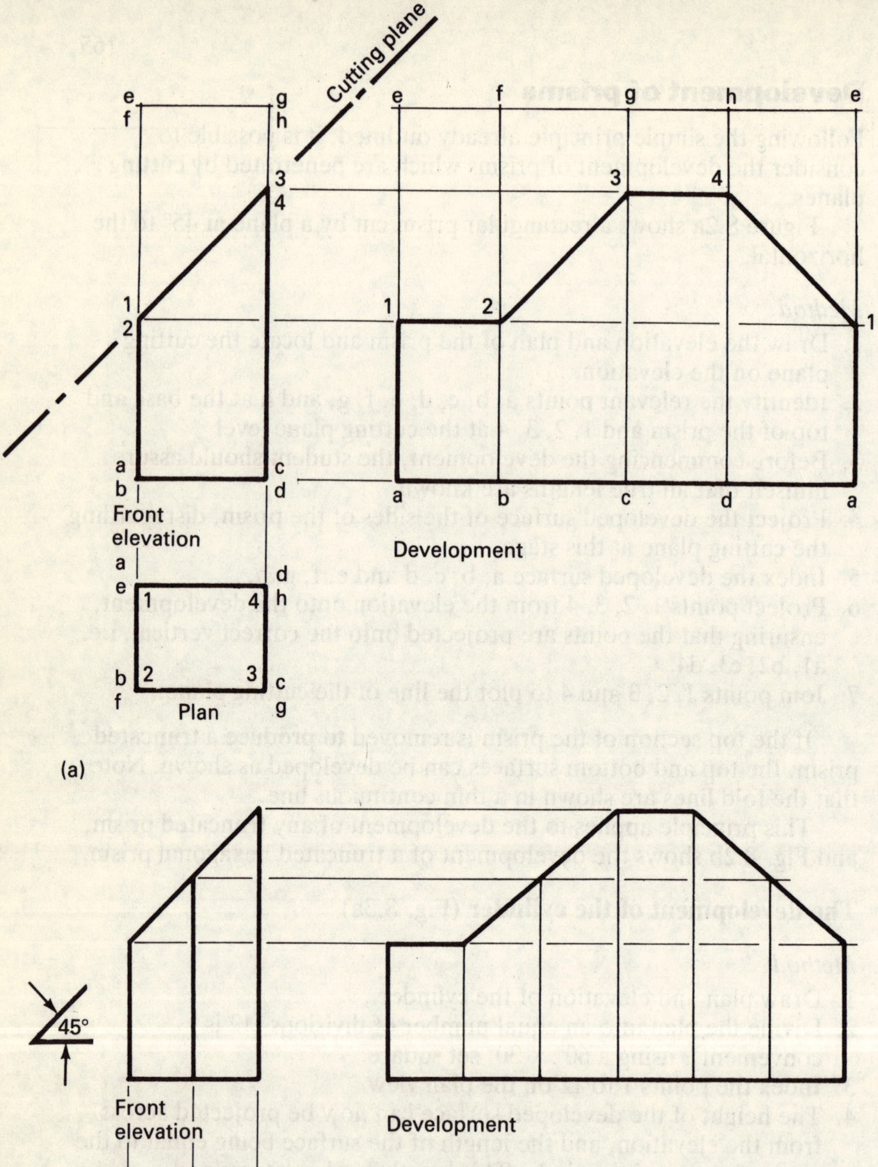

Fig. 8.2 Development of prisms

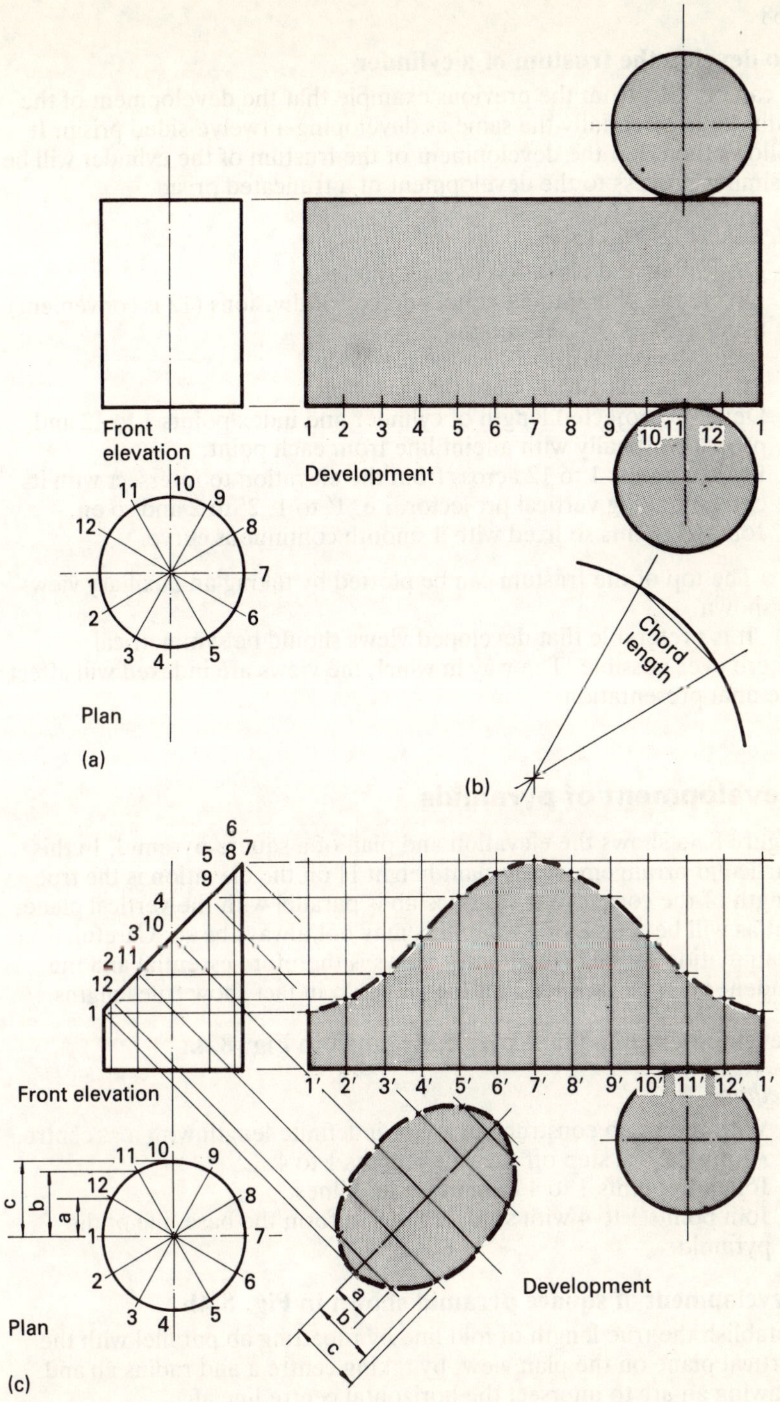
Front elevation
Development
Plan
(a)
Chord length
(b)
Front elevation
Development
Plan
(c)
1 2 3 4 5 6 7 8 9 10 11 12 1
1′ 2′ 3′ 4′ 5′ 6′ 7′ 8′ 9′ 10′ 11′ 12′ 1′
a
b
c

Fig. 8.3

To develop the frustum of a cylinder

It can be seen from the previous example that the development of the cylinder is essentially the same as developing a twelve-sided prism. It follows then that the development of the frustum of the cylinder will be a similar process to the development of a truncated prism.

Method (Fig. 8.3c)

1. Draw plan and elevation of the cylinder.
2. Divide the plan into an equal number of divisions (12 is convenient) using a 60° × 30° set square.
3. Index the points 1 to 12 on the plan view.
4. Project points 1 to 12 onto the elevation.
5. Develop projected length of cylinder and index points 1 to 12 and project vertically with a feint line from each point.
6. Project points 1 to 12 across from the elevation to intersect with its corresponding vertical projector, i.e. 1′ to 1, 2′ to 2 and so on.
7. Join the points so fixed with a smooth continuous curve.

The top of the frustum can be plotted by taking an auxiliary view as shown.

It is preferable that developed views should be symmetrical where ever possible. The way in which the views are indexed will affect the final presentation.

Development of pyramids

Figure 8.4a shows the elevation and plan of a square pyramid. In this particular arrangement the slant height H on the elevation is the true length of the corner fold ab, since ab is parallel with the vertical plane, but as will be seen in Fig. 8.4b this may not always be so. Careful examination of the orthographic views is therefore essential and the student must be satisfied that the views do in fact show true lengths.

Development of square pyramid shown in Fig. 8.4a

Method

1. With radius ab construct an arc of indefinite length with a as centre.
2. Along the arc step off the side lengths 1 to 4.
3. Join a to points 1 to 4 to indicate fold lines.
4. Join points 1 to 4 with straight lines to form the base line of the pyramid.

Development of square pyramid shown in Fig. 8.4b

Establish the true length of fold line by rabatting ab parallel with the vertical plane on the plan view, by taking centre a and radius ab and drawing an arc to intersect the horizontal centre line afc.

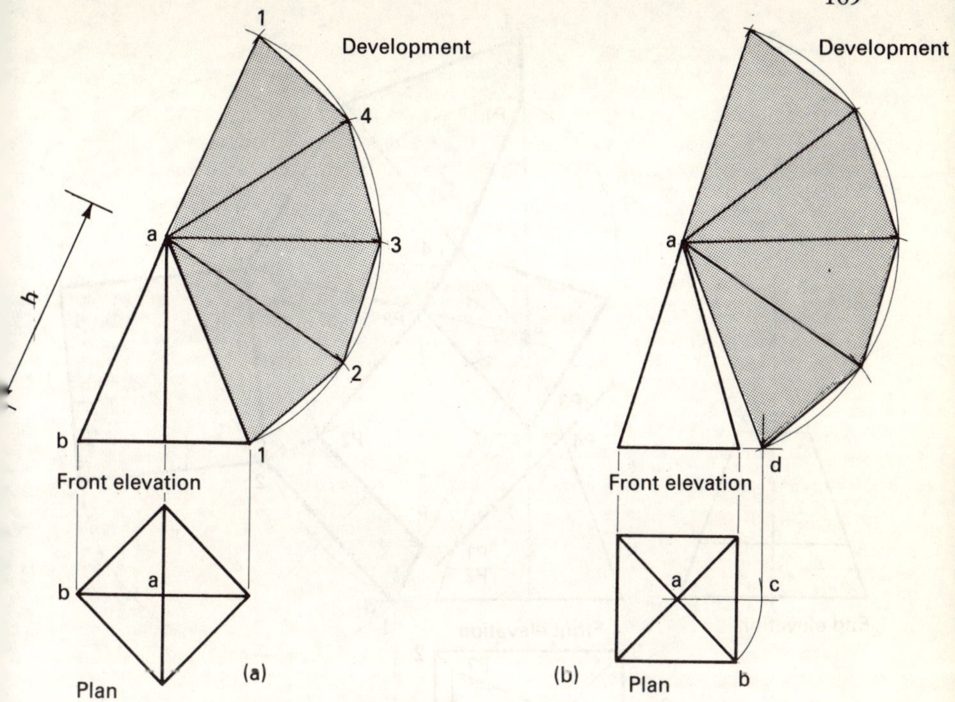

Fig. 8.4 Development of Pyramids

Project c to d on the extended base of the pyramid seen on the elevation and join d to a, da is the true length of the corner fold.

The development is carried out as the previous example using a radius ad to begin the development.

By comparison it will be seen that in both of the examples the final developments are identical.

Development of the frustum of a pyramid

The square pyramid shown in Fig. 8.5 is cut by an inclined plane as shown.

It is required to develop the surface of the pyramid below the cutting plane.

Method

1. Develop initially the surface of the entire pyramid as previously described.
2. The elevation shows the plane cutting the corner fold lines at P1, P2, P3 and P4 and these points are projected across onto the true length line ad.

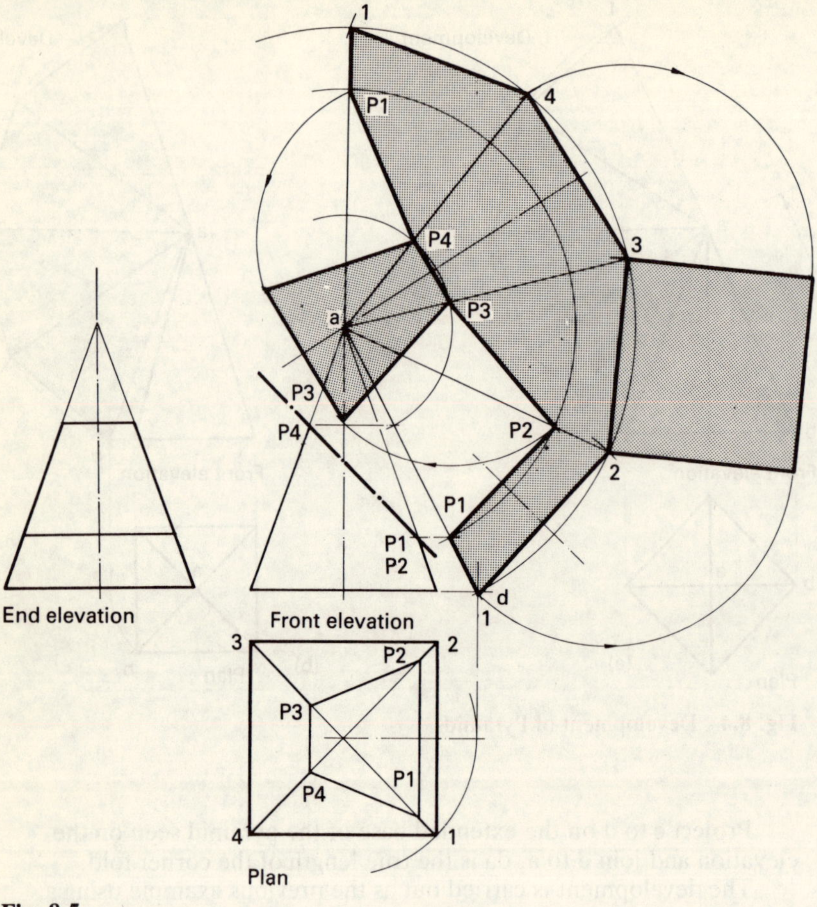

Fig. 8.5

3. With centre a and appropriate radii construct arcs to give intersection P 1.1, P 2.2, P 3.3 and P 4.4.
4. Join the points with straight lines to produce the required development.

Development of the cone

Figure 8.6a shows the elevation of a cone of diameter D and slant height H. It is required to develop the conical surface.

Method

1. With radius H construct an arc of indefinite length, with the apex of the cone as centre.

2. Divide the plan view into a convenient number of divisions (12 shown) and index 1 to 12.
3. Step off the divisions 1 to 12 along the developed arc to determine the developed length of the base of the cone. Index the points around the developed arc.
4. Join point 1′ back to the apex of the cone to give the required development.

Development of the frustum of a cone

Figure 8.6b shows a cone cut by an inclined plane. The portion of the cone below the cutting plane is known as the frustum, and it is the development of the frustum which is required.

Method

1. With radius H construct an arc of indefinite length, with the apex of the cone as centre.
2. Divide the plan view into a convenient number of divisions (12 shown) and index 1 to 12.
3. Step off the divisions 1 to 12 along the developed arc to determine the developed length of the base of the cone, index the points around the developed arc and join points back to apex.
4. Project points 1 to 12 onto the elevation, and join the intersections at the base to the apex of the cone, using generators.
5. The points of intersection between the generators and the cutting plane are projected across onto the slant height p1, p2, p3, etc.
6. With apex as centre construct arcs of radius p1, p2, p3, etc. to intersect with the respective generators on the developed surface, i.e. 1.1′, 2.2′, 3.3′.
7. Join the points plotted with a smooth curve to produce the required development.

Development of roof surface

By adopting the principles already established, it is possible to develop the surface of a roof, which may prove particularly useful in determining the true length of hip and valley rafters.

Figure 8.7 shows the elevation and plan of a hipped roof. It is required to find the true length of the hip rafter and to develop the surface of the roof.

Method

1. With centre a on the elevation rabat the slant height of the roof parallel to the horizontal plane, and project onto plan to intersect at g. Join g to a and d to produce development of the hipped end.
2. ga and gd represent the true length of the hip rafter.

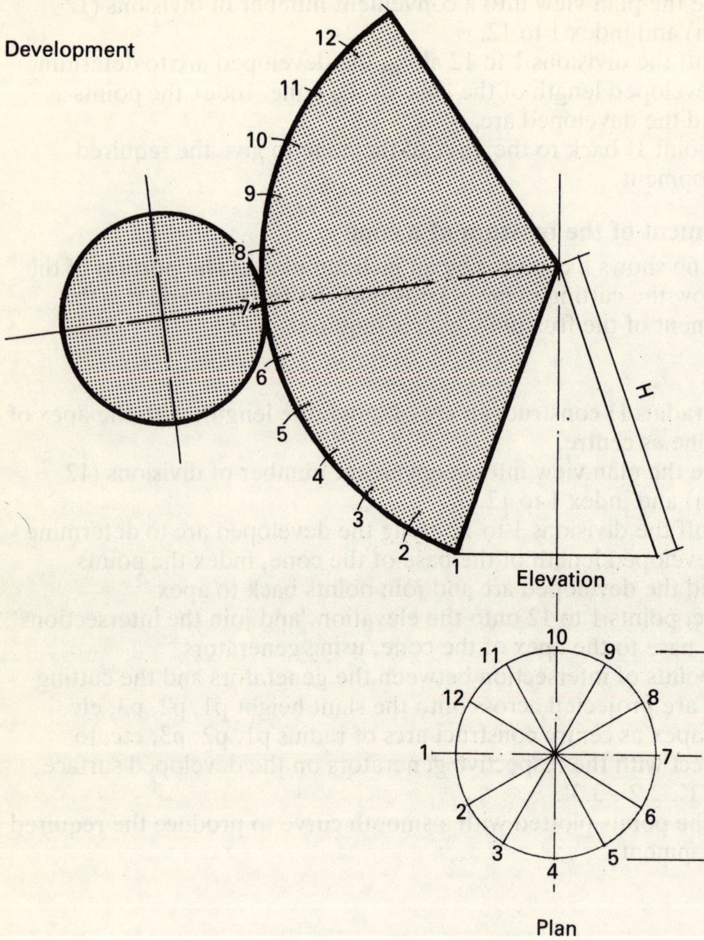

Fig. 8.6(a) Development of the surface of a cone

3. With centre a on the plan draw an arc to intersect the perpendicular from e at h (ah also represents the true length of the hip rafter).
4. Repeat for opposite end of roof to produce point j.
5. Join hj to produce development of roof side.
6. Repeat the process for the remaining roof slopes to complete the full development.

If it is required only to determine the length of the hip rafter, this can be found by a similar method to that used in the development of pyramids, and shown on the right-hand end of Fig. 8.7.

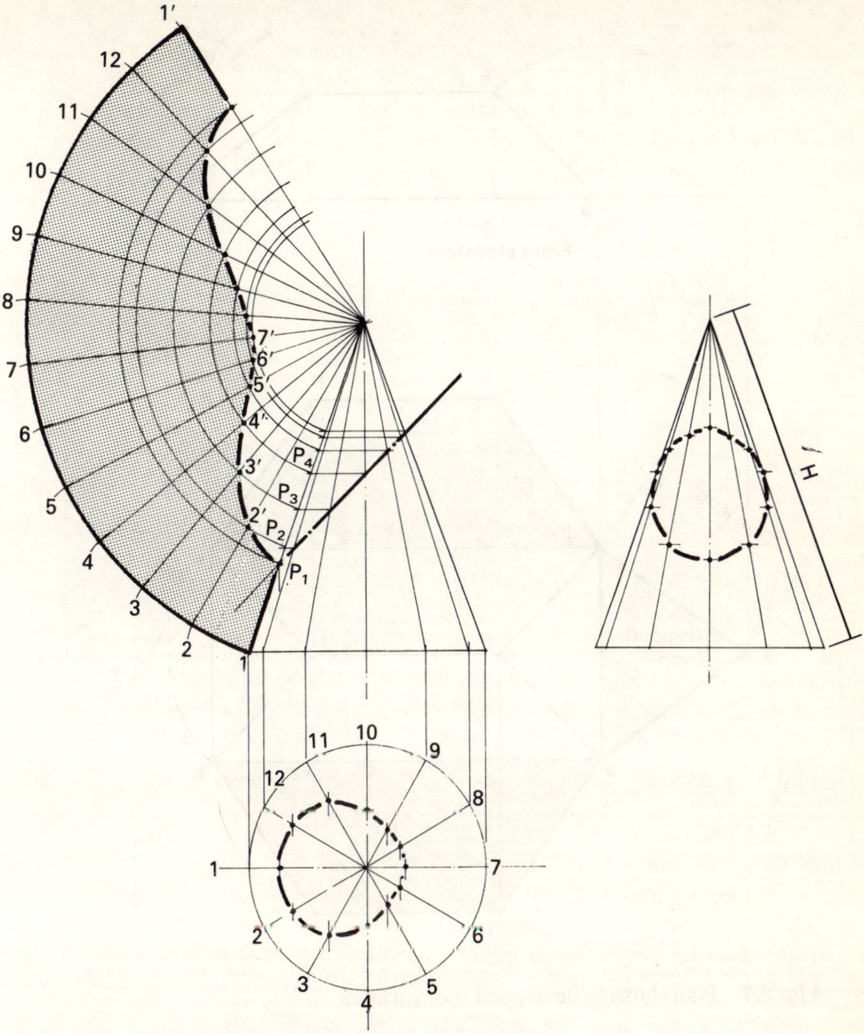

Fig. 8.6(b) Development of the frustum of a cone

Method

1. With centre f and radius fb, construct an arc to intersect the extended ridge line at K.
2. Project vertically from K onto the front elevation to intersect the extended eaves line at m.
3. Join m to f, giving the true length of the hip.

Figure 8.8 shows the application of roof development to a small dormer roof.

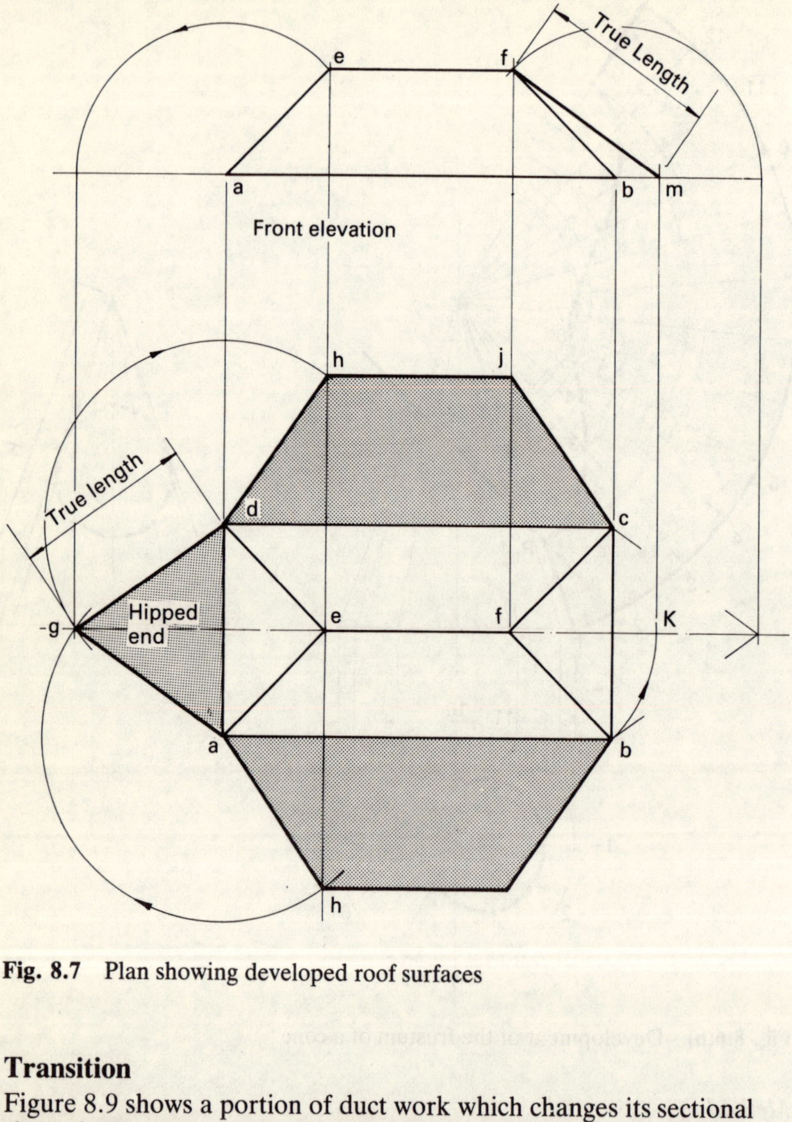

Fig. 8.7 Plan showing developed roof surfaces

Transition

Figure 8.9 shows a portion of duct work which changes its sectional shape from square to circular. Such changes in shape are known as transition pieces and are frequently encountered in the drawing of ducting, chimney hoods and hoppers.

It is required to develop the surface of the transition piece.

Method

1. Draw the front elevation and plan of the piece. It will be appreciated that on the plan view the square base and circular top are seen as true lengths and true shapes.

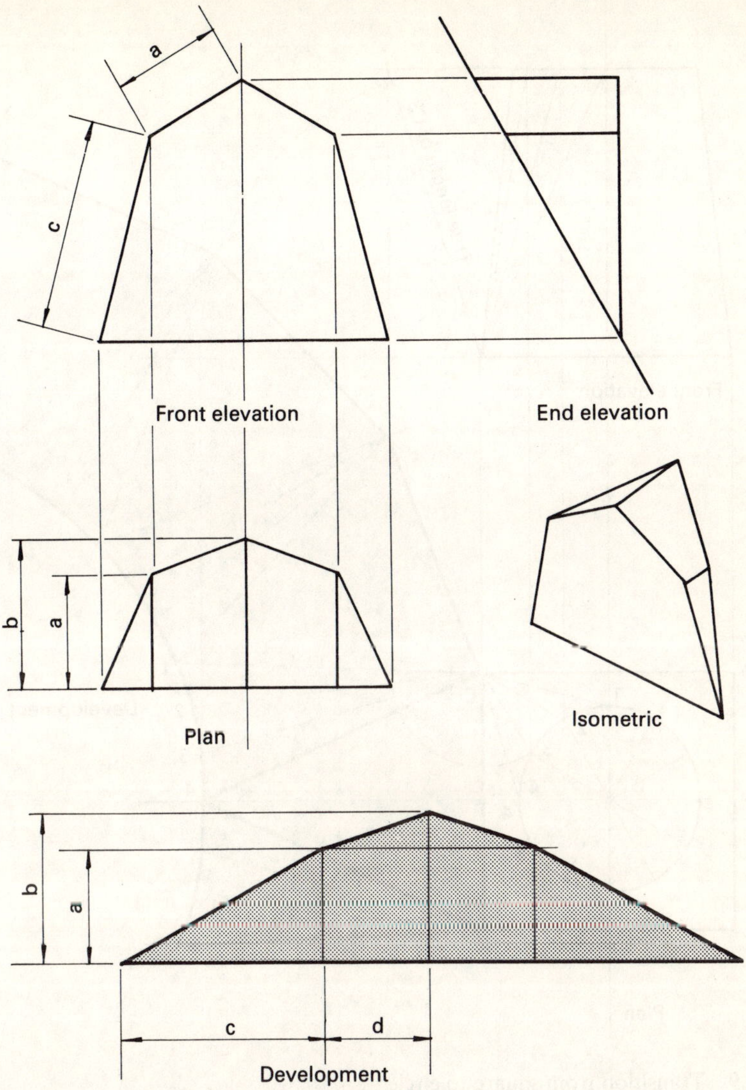

Fig. 8.8 Development of a dormer roof

2. Divide the circle into 12 and join points to the corners of the base at ABC and D.
3. Determine the true height of triangle CB4 by rabatment and construct the true shape CB4′ on the base CB.
4. Determine the true length of C3 by rabatment. Note that C1 = C4 and C2 = C3 and these in turn are equal to their counterparts in the remaining three quadrants.

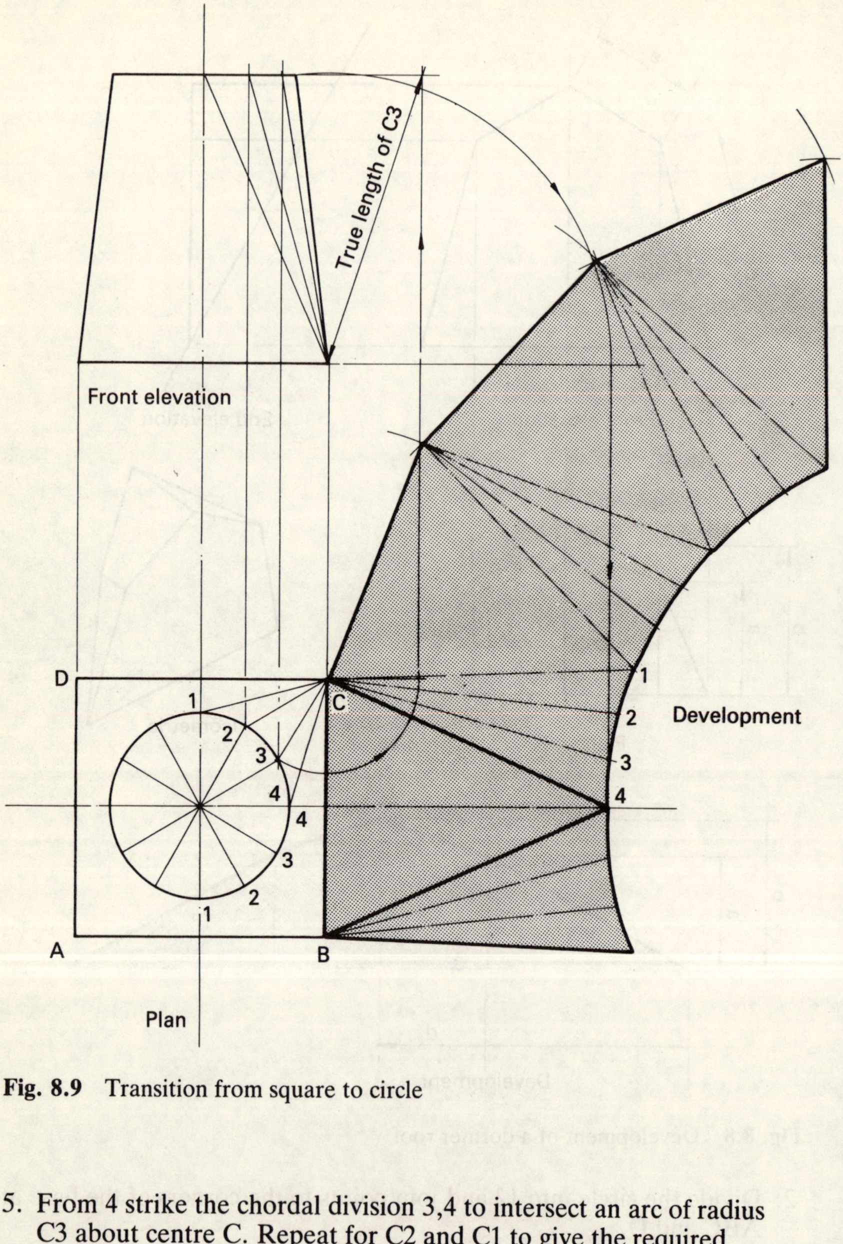

Fig. 8.9 Transition from square to circle

5. From 4 strike the chordal division 3,4 to intersect an arc of radius C3 about centre C. Repeat for C2 and C1 to give the required points.
6. Join the points plotted to produce the required development. This method may be utilized to develop other transitions such as rectangles to circles, and square to ellipse, etc.

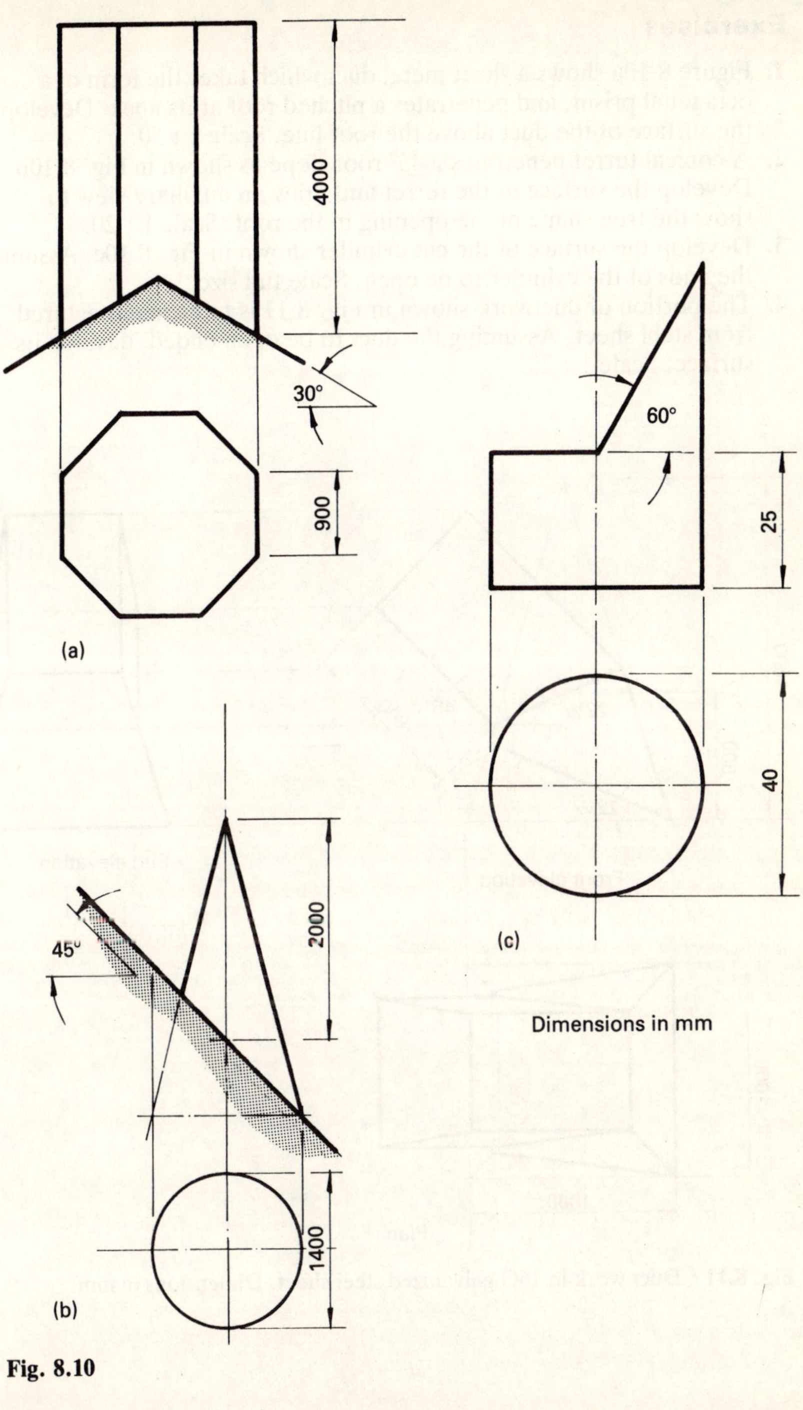

Fig. 8.10

Exercises

1. Figure 8.10a shows a sheet metal duct which takes the form of a octagonal prism, and penetrates a pitched roof at its apex. Develop the surface of the duct above the roof line. Scale 1 : 50.
2. A conical turret penetrates a 45° roof slope as shown in Fig. 8.10b. Develop the surface of the turret and draw an auxiliary view to show the true shape of the opening in the roof. Scale 1 : 20.
3. Develop the surface of the cut cylinder shown in Fig. 8.10c. Assume the ends of the cylinder to be open. Scale full size.
4. The portion of ductwork shown in Fig. 8.11 is to be manufactured from steel sheet. Assuming the duct to be open ended, develop its surface. Scale 1 : 20.

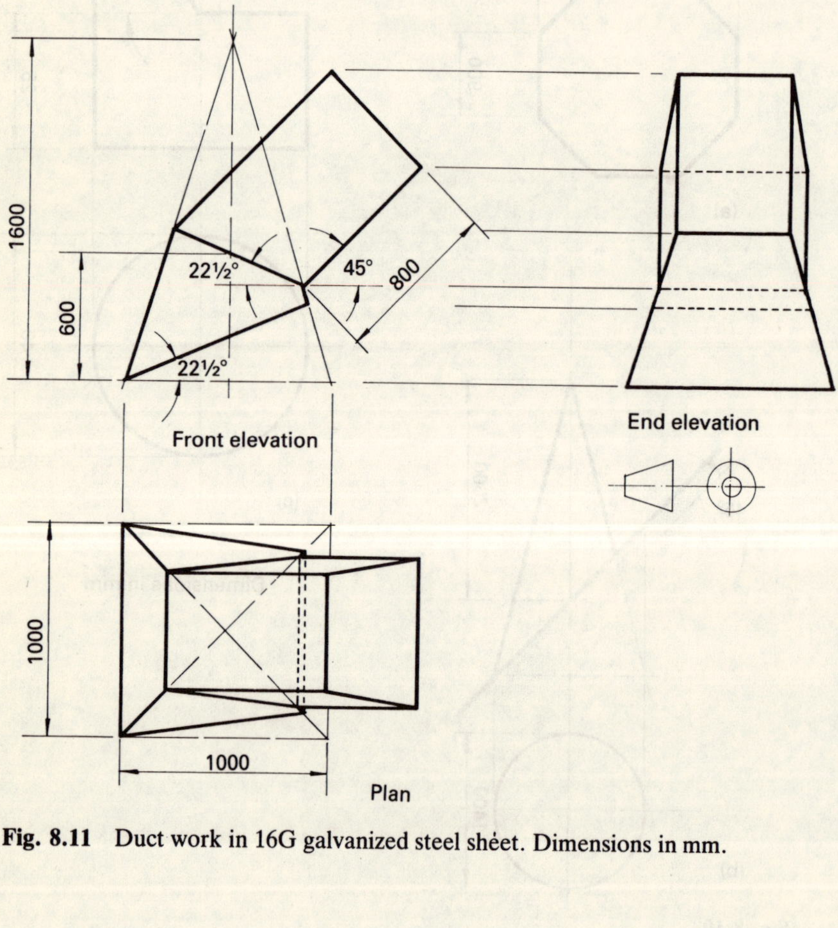

Fig. 8.11 Duct work in 16G galvanized steel sheet. Dimensions in mm.

Chapter 9

The graphical solution of beams

Before dealing with the application of graphical methods to beam design, it is essential that the student has a clear understanding of the basic terminology used in connection with the study of forces.

The following terms must firstly be considered:

Co-planar forces. Forces which act in the same single plane.
Concurrent forces. A system of two or more forces, having lines of action which pass through a common point (Fig. 9.1a).
Non-concurrent forces. A system of two or more forces having lines of action with no common point of concurrence (Fig. 9.1b).
Parallel forces. A system of two or more forces having parallel lines of action. These may either be like or unlike parallel systems (Fig. 9.1c and d). In dealing with the solution of simple beams, parallel force systems predominate, and particular emphasis will be given to this case.
Resultant of a system of forces. The resultant of a given system of forces is the single force which can replace the system and maintain the same effect upon the body upon which the forces act (Fig. 9.1e).

For a beam to remain in equilibrium the resultant of all the forces acting upon it must equal zero.

Equilibrant of a system of forces. The equilibrant of a system of forces is the single force which, when introduced into a force system, maintains the system in a state of equilibrium (Fig. 9.1e). It will be equal in magnitude but opposite in direction to the resultant.

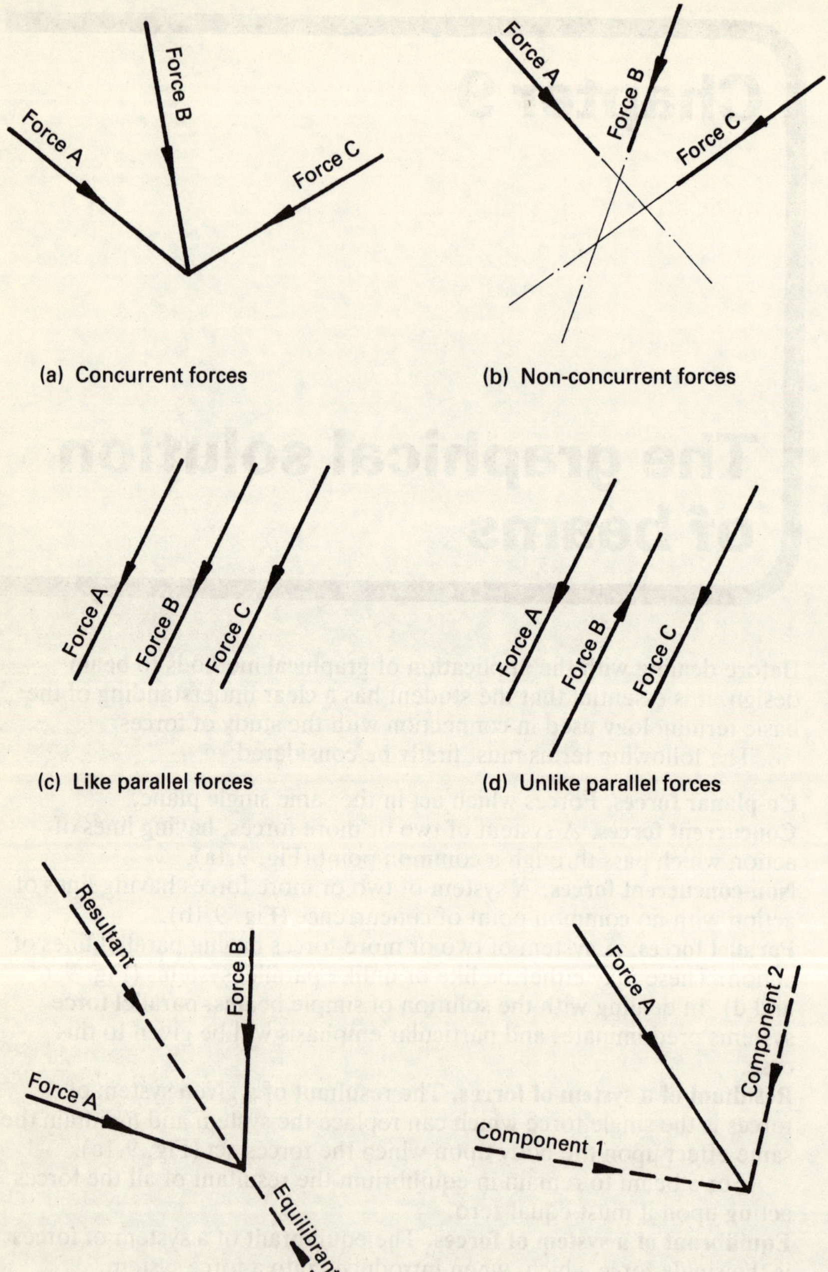

Fig. 9.1 Forces and force systems

Component. The components of a single force are the separate forces which can replace the given force without causing change to the effect of the force. The separate forces will have the original single force as their resultant (Fig. 9.1f).

Graphical representation of a force

The properties of a co-planar force can be illustrated accurately by a single line known as a vector line.

The properties of the force which need to be illustrated are:

1. **Magnitude.** Its value, i.e. 5 kN, 10 kN, etc.
2. **Direction.** Vertically downward, horizontal, etc.
3. **Position.** Its relationship to other loads, supports, etc.

If all three of these properties are to be accurately conveyed it is necessary to give careful consideration to scales. The vector line length will represent the magnitude of this force; the scale must therefore relate length to force; for example, 10 mm represents 1 kN or 5 mm represents 2 kN depending upon the range of forces being dealt with. It follows that the importance of clearly stating the scale used cannot be over emphasized.

It will be seen later in the chapter that force scale and linear scale are used simultaneously, and a clear distinction must be made between the two to avoid confusion.

The direction of the force is represented by the direction of the vector line, and it is a fundemental essential that a vector line must be drawn parallel to the line of action of the force it represents. The 'sense' of the force is indicated by an appropriate arrowhead. A line which indicates both the magnitude and direction of a force is known as a vector quantity (Fig. 9.2a). The position of a force can be shown graphically simply by drawing to scale the point of application relative to other parts of the structure or by showing its relationship to other forces in a system, where perhaps several forces might be acting on a point. The positioning of a force will often necessitate the use of linear scale and again the scale used must be clearly stated (Fig. 9.2b).

Figure 9.2c shows an example of a force of 10 kN acting at a given point P vertically downward. Note the arrow indicating the sense of the force and also the clear indication of the scale used in its construction.

Identification of forces

When a system of forces is shown in the space diagram, it is important to clearly identify each individual force in the system. To do this a standard approach is adopted known as BOW'S NOTATION. The correct application of Bow's Notation is essential in graphical solutions and the student must be familiar with its use.

To identify the forces in a system a capital letter is placed in the space between the lines of action of each force.

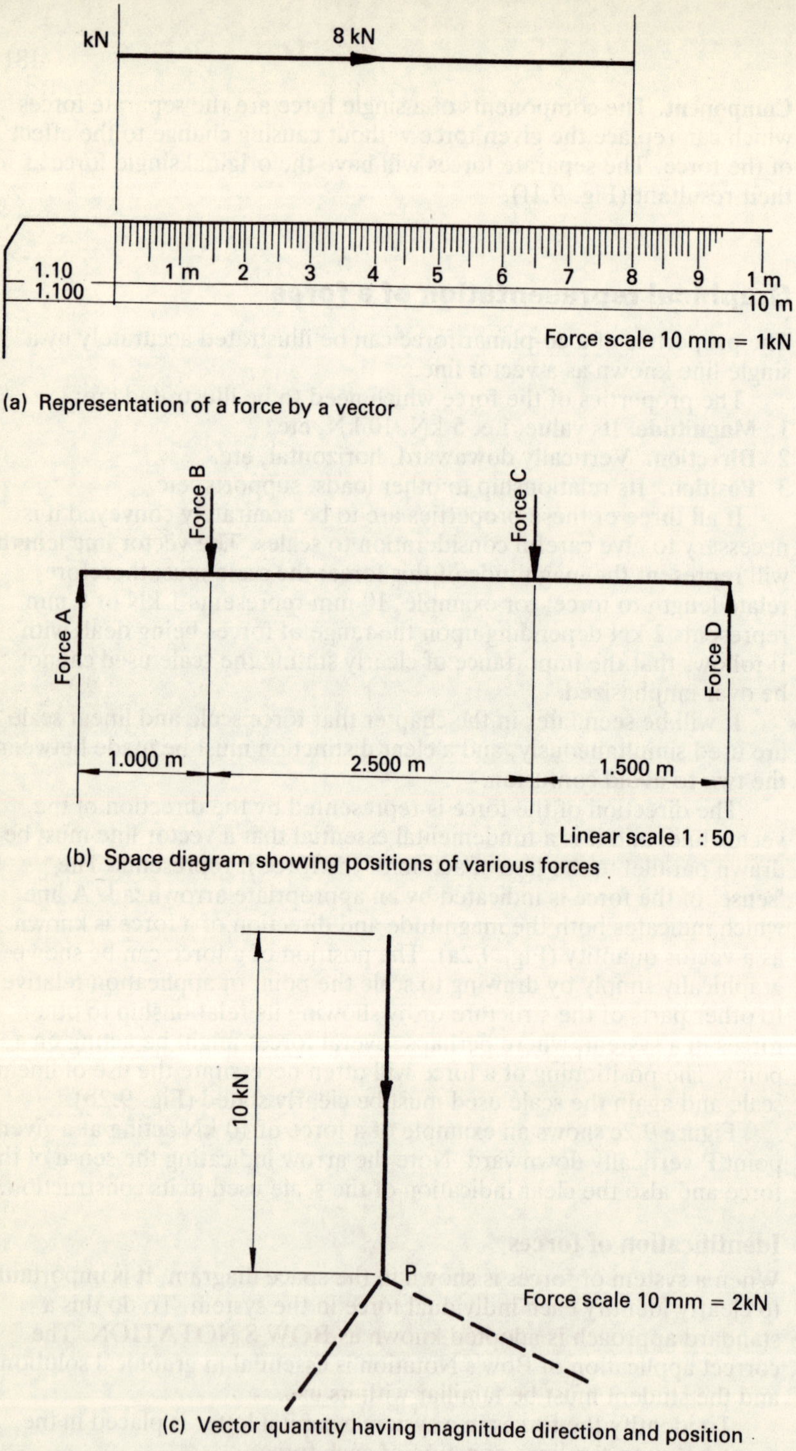

(a) Representation of a force by a vector

(b) Space diagram showing positions of various forces

(c) Vector quantity having magnitude direction and position

Fig. 9.2 Representation of forces

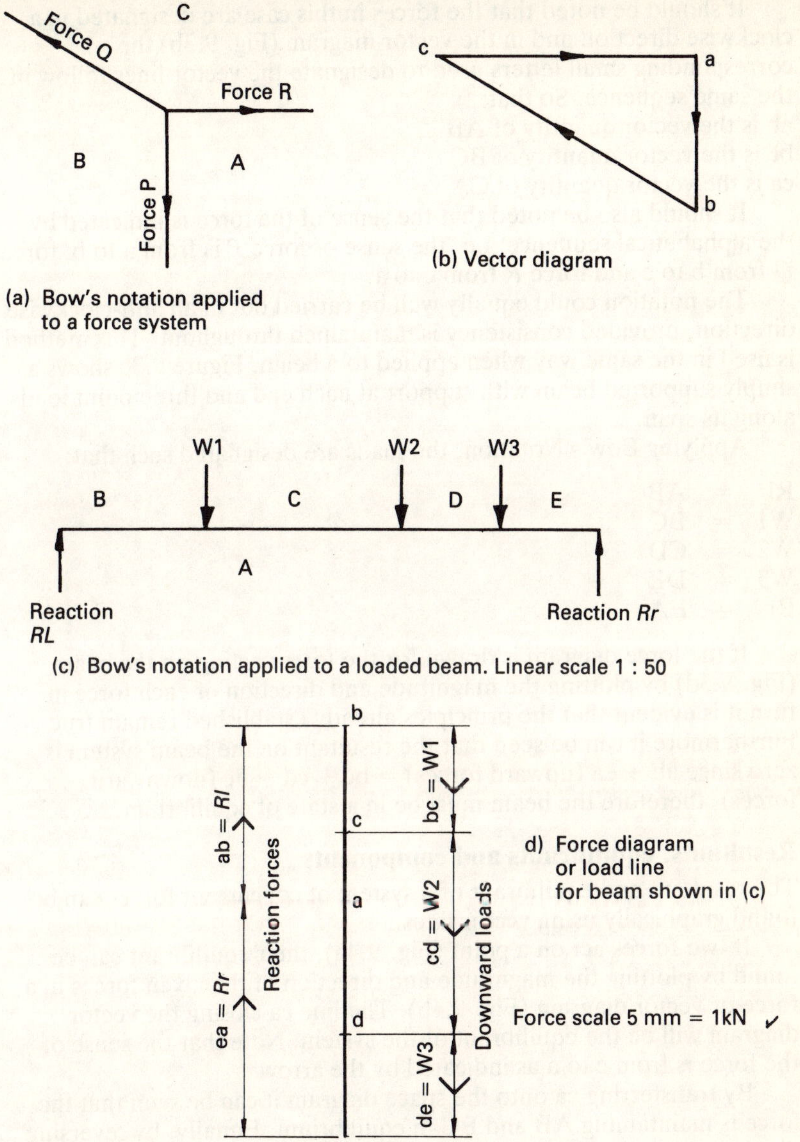

Fig. 9.3 Identification of forces

Each force is then designated by the capital letter positioned either side of the line of action of the force.

It can be seen then by reference to Fig. 9.3a that:

Force *P* becomes AB
Force *Q* becomes BC
Force *R* becomes CA

It should be noted that the forces in this case are designated in a clockwise direction and in the vector diagram (Fig. 9.3b) the corresponding small letters used to designate the vector lines follow in the same sequence. So that:
ab is the vector quantity of AB
bc is the vector quantity of BC
ca is the vector quantity of CA

It should also be noted that the sense of the force is indicated by the alphabetical sequence, i.e. the sense of force *P* is from a to b, force *Q* from b to c and force *R* from c to a.

The notation could equally well be carried out in an anti-clockwise direction, provided consistency is maintained throughout. This method is used in the same way when applied to a beam. Figure 9.3c shows a simply supported beam with support at each end and three point loads along its span.

Applying Bow's Notation, the loads are designated such that:

Rl	=	AB
W1	=	BC
W2	=	CD
W3	=	DE
Rr	=	EA

If the force diagram is drawn for the forces acting on the beam (Fig. 9.3d) by plotting the magnitude and direction of each force in turn it is evident that the principles already established remain true. Furthermore it can be seen that the resultant on the beam system is zero since ab + ea (upward forces) = bc + cd + de (downward forces), therefore the beam must be in a state of equilibrium.

Resultants, equilibrants and components

The resultant and equilibrant of a system of concurrent forces can be found graphically using vector lines.

If two forces act on a point (Fig. 9.4a), their equilibrant can be found by plotting the magnitude and direction of the given forces in a force or vector diagram (Fig. 9.4b). The line ca closing the vector diagram will be the equilibrant of the system. Note that the sense of the force is from c to a as indicated by the arrow.

By transferring ca onto the space diagram it can be seen that the force is maintaining AB and BC in equilibrium. Equally, by reversing the direction of the force to become ac (Fig. 9.4c), it is evident that CA becomes the resultant of AB and BC (resultant is equal in magnitude and opposite in direction to the equilibrant). It should be noted, then, that in a force diagram the arrowhead representing the resultant vector line is reversed.

Where more than two forces act on a point the same principles may be applied. Figure 9.4d shows four forces acting on a point X.

A force diagram is constructed as before and the diagram closed by

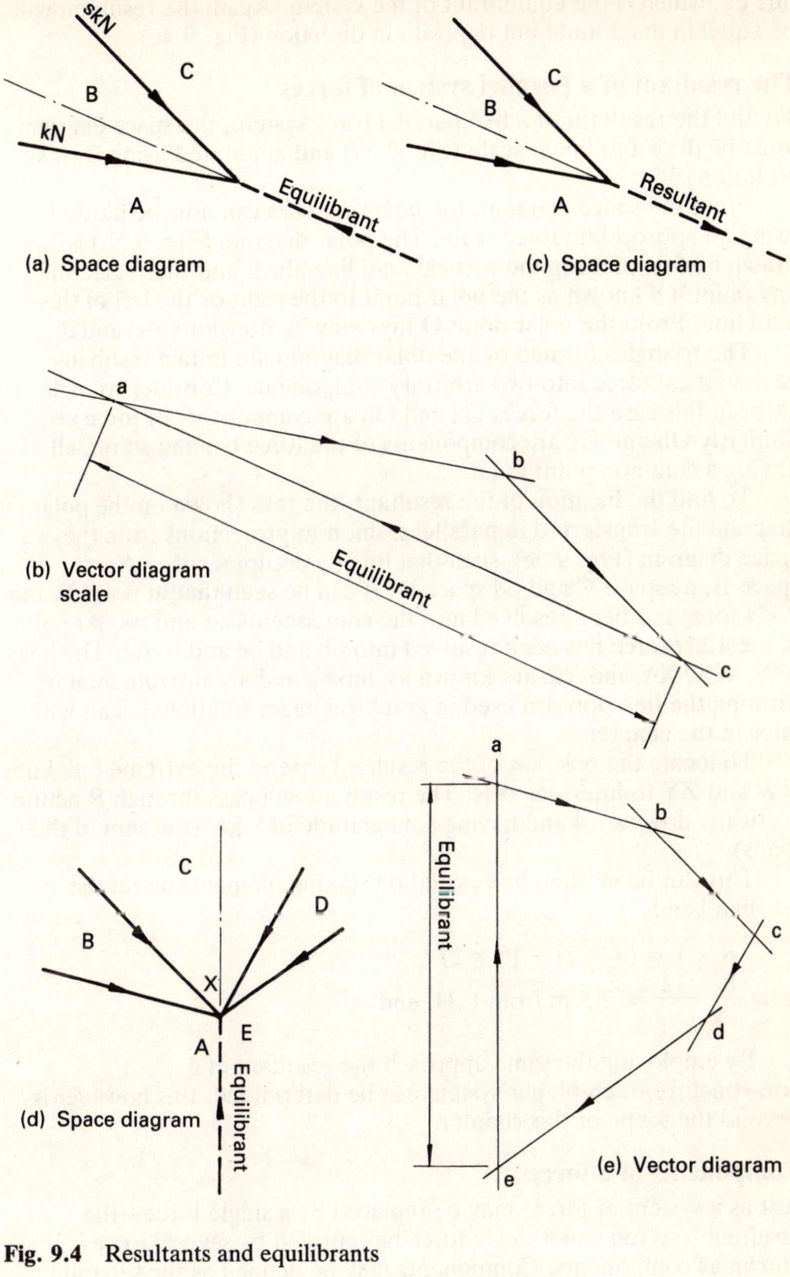

Fig. 9.4 Resultants and equilibrants

line ea, which is the equilibrant of the system. Again the resultant will be equal in magnitude but opposite in direction (Fig. 9.4e).

The resultant of a parallel system of forces

To find the resultant of a like parallel force system, the space diagram must be drawn to linear scale (Fig. 9.5a) and annotated using Bow's Notation.

From the space diagram, the *polar diagram* can now be plotted using an appropriate force scale. The polar diagram (Fig. 9.5b) is drawn by first plotting the vertical load line abcd, and then selecting any point 'O' known as the polar point to the right or the left of the load line. From the polar point O rays may be drawn to abc and d.

Thc triangles formed by the polar diagram are in fact resolving each vertical force into two arbitrary components. Consider triangle Oab. In this case the forces Oa and Ob are components of force ab. Similarly Ob and Oc are components of the force bc, and so on, all having a common point in O.

To find the location of the resultant, the rays shown on the polar diagram are transferred in parallel fashion to projections from the space diagram (Fig. 9.5c), such that line oa occupies space A, ob space B, oc space C and od space D. It can be seen that at point W the 2 kN force has been resolved into the components ao and bo. At point X the 4 kN force has been resolved into ob and oc and so on. The lines VW, WX, XY and YZ are known as 'links', and are instrumental in forming the link polygon used in graphical beam solutions, dealt with later in the chapter.

To locate the position of the resultant extend the extreme link lines VW and ZY to intersect at R. The resultant will pass through R acting vertically downward and having a magnitude of 8 kN (the sum of the loads).

This can be verified by calculation, taking moments about the left-hand end.

$$8 \times x = (4 \times 2) + (6 \times 2)$$

$$x = \frac{8 + 12}{8} = 2.5 \text{ m from L.H. end}$$

By employing the same approach the resultant of a non-concurrent, co-planar system can be determined; this however is beyond the scope of this chapter.

Components of a force

Just as a system of forces may be replaced by a single force – the resultant – so too can a single force be replaced by several forces known as components. Components may be defined as the separate forces, which acting all together will produce the same force as their resultant.

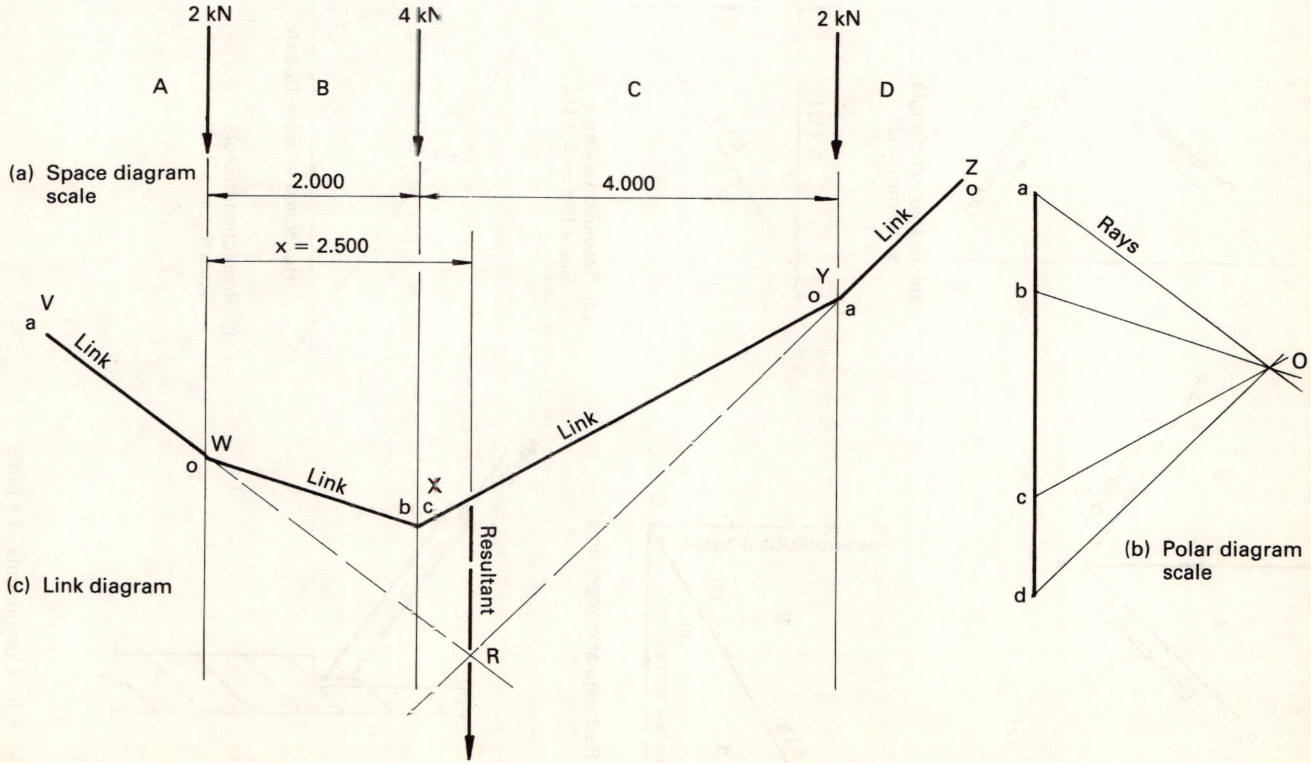

Fig. 9.5 Resultant of a parallel system of forces

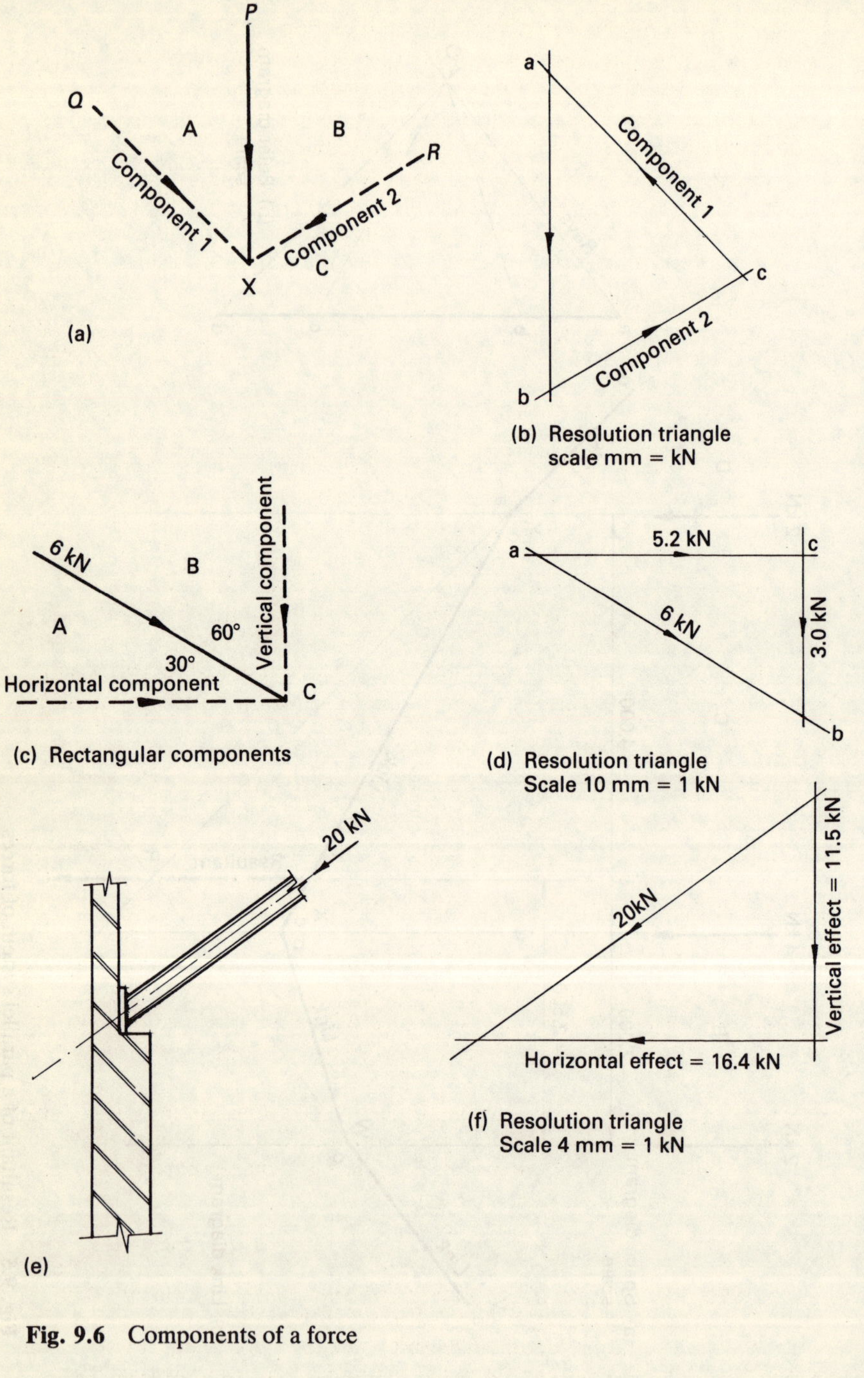

Fig. 9.6 Components of a force

It is often necessary in structural calculation to ‘break down’ a force into particular components, an operation known as resolution; a force is then said to be resolved into its various components.

Although a force can have an infinite number of components it is not often necessary to resolve into more than two.

Figure 9.6a shows a single force P acting vertically downward. This force could be removed from its point of application X and replaced by the two forces Q and R without any change to the effect on point X. Forces Q and R are components of P, and are shown in the force diagram (Fig. 9.6b).

Often in a design problem, particular components are required; for example, the vertical and horizontal components of a force acting at an angle to the horizontal (Fig. 9.6c). The force of 6 kN can be resolved into the two components again using vector lines (Fig. 9.6d). ab is plotted parallel to the line of action of the force, from b a vertical line is drawn to intersect the horizontal from a at c. ac and cb represent the components of ab, note the directions of the arrows in the force diagram. It can be seen that ab is the resultant of ac and cb.

Figure 9.6e shows a wall subject to a thrust of 20 kN from a strut. The resolution triangle determines the horizontal thrust the wall must resist and also the vertical load which will be transmitted to the foundation.

It will be seen later in the chapter that for graphical beam design the process of resolution is arbitrary and is simply a means to an end.

Beam reactions

The values of beam reactions may be resolved graphically by application of the polar diagram.

As described earlier in the chapter, the polar diagram is a series of vector triangles which resolve the vertical loads on the beam into arbitrary components which concur at the polar point.

In order for a beam to maintain a state of equilibrium, the reaction forces must equal the downward force, and it follows therefore that the total reaction magnitude of the load line, drawn to scale, will equal the length of the load line plotted in the initial stages of drawing the polar diagram. The sense of the reaction forces will of course be opposite to the sense of the load.

At the outset then it is understood that the magnitude of the total support reactions will equal the total downward loads, but the polar diagram used in conjunction with the *link polygon* enables the individual reaction values to be determined.

Method

To a suitable linear scale draw the space diagram for the beam, and annotate the beam using Bow’s Notation (Fig. 9.7a).

Select a suitable vector scale and draw the polar diagram (Fig. 9.7b). Project down from the loading diagram, and construct

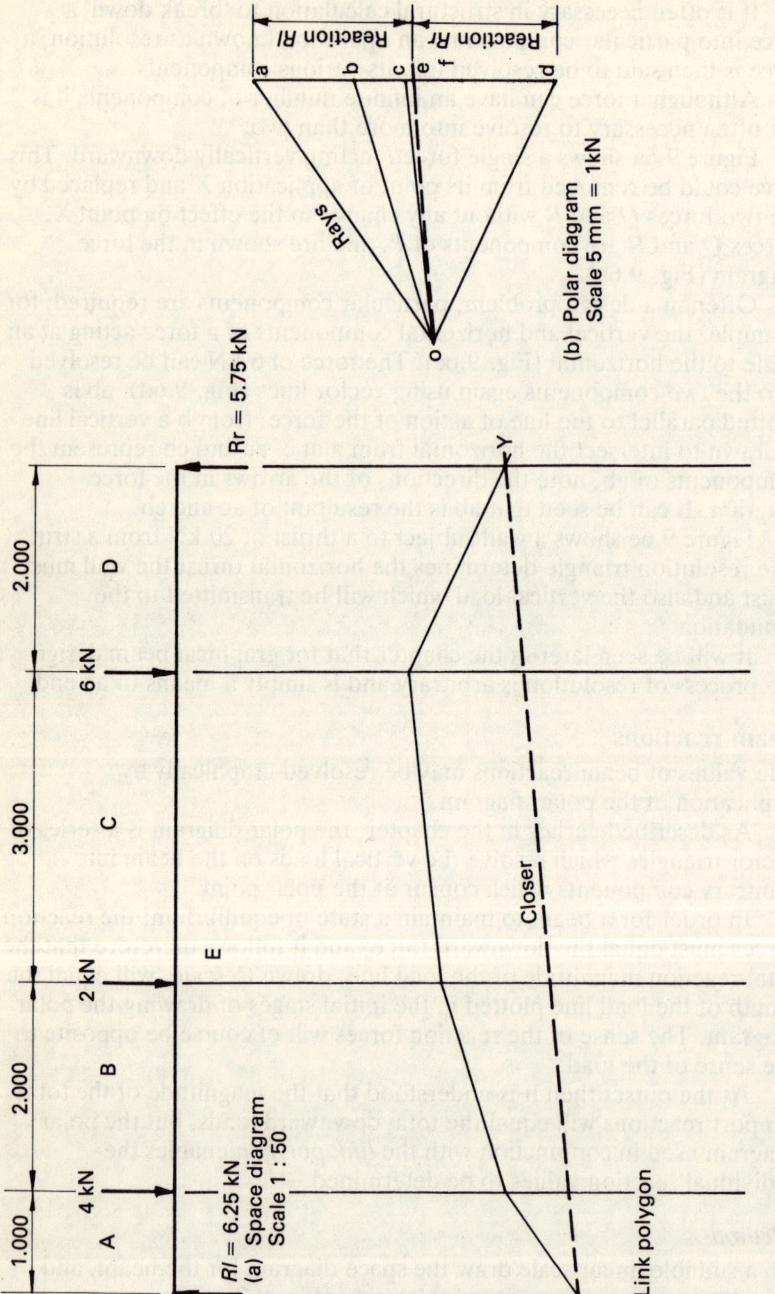

Fig. 9.7 Beam reactions by graphical method

beneath it the link polygon by extending the links at the outer extremes to intersect with the reaction lines Rl and Rr. (Figure 9.7c) The points of intersection X and Y are joined by a line called a *closer* forming the polygon. The closer line is then transferred in parallel fashion to the polar diagram such that it passes through the pole O, and intersects the load line at e. The vector ea will then represent the magnitude of Rl and de the magnitude of Rr. Note that the forces are given sense by their notation e to a and d to e.

Shear force

The shear force diagram can be projected directly from the polar diagram, since the vertical steps in the shear force diagram represent the beam loads in the same way as the load line in the polar diagram. The shear force diagram will in this case be drawn to the same force scale as the load line on the polar diagram.

The base or datum line for the shear force diagram is projected across from the point of intersection e between the closer and the load line on the polar diagram, the vertical distance ea representing the left-hand reaction and de the right-hand reaction. By now projecting the points abc and d across to intersect with their respective lines of action from the space diagram the required points can be plotted (Fig. 9.8).

This again can be verified by simple calculation, remembering that the shear force is the resultant vertical force of all the forces acting to one side of the section.

Consider shear force at point XY 1.5 m from Rl.

Resultant forces to the left = 6.25 – 4   2.25.

furthermore,

Resultant of forces to the right = 5.75 – (6 + 2) = 2.25.

Bending moment

The bending moment diagram may be considered as a graph showing the variations in the values of the bending moments occurring along the length of a beam. The link polygon previously drawn does to some undetermined scale represent the bending moment diagram, and for this reason some consideration should be given to its presentation in that it will represent positive bending and negative bending; positive bending being indicated above the datum and negative below.

In order to maintain recognized convention the polar point must be positioned to the left of the load line, such that when the link polygon is formed positive bending will be above datum and negative below.

Having constructed the closer forming the link polygon, the diagram can now be re-plotted on a horizontal base line (Fig. 9.9), ensuring when doing so that vertical distances are accurately maintained.

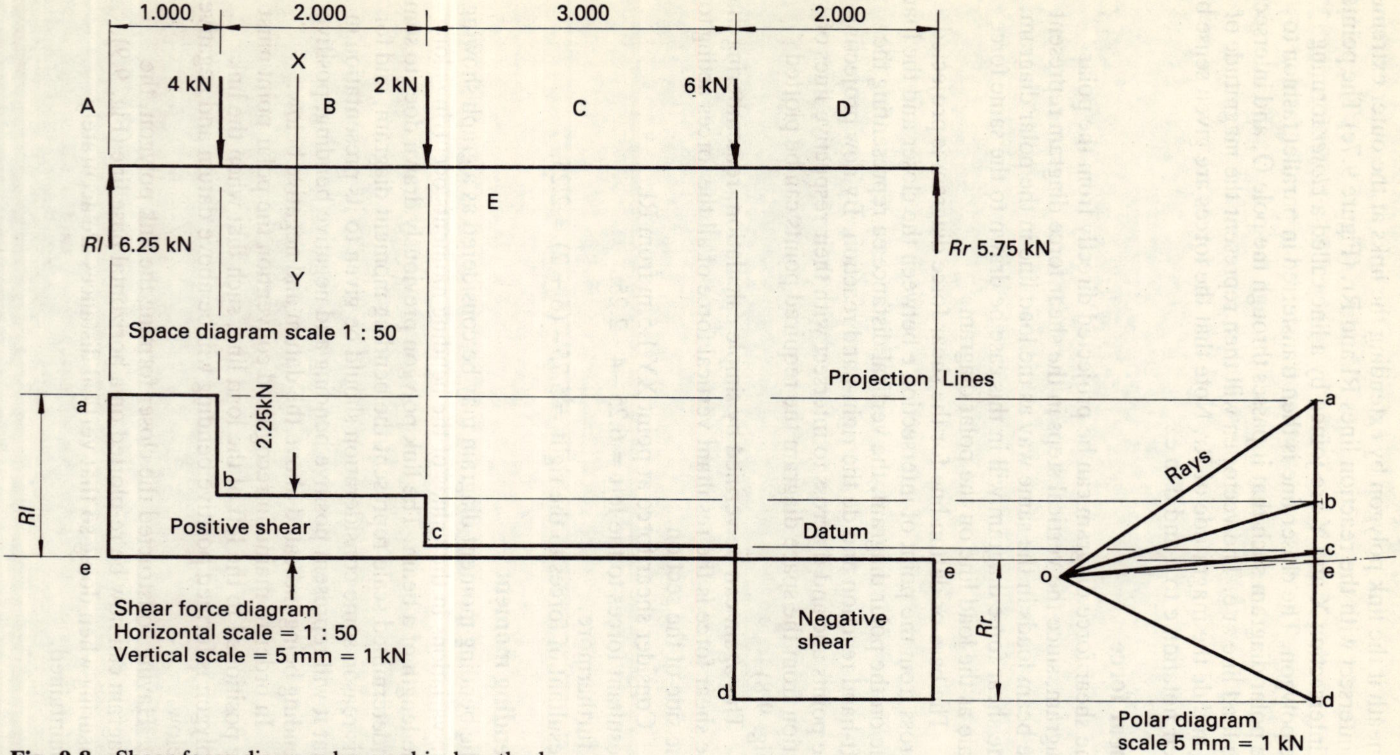

Fig. 9.8 Shear force diagram by graphical method

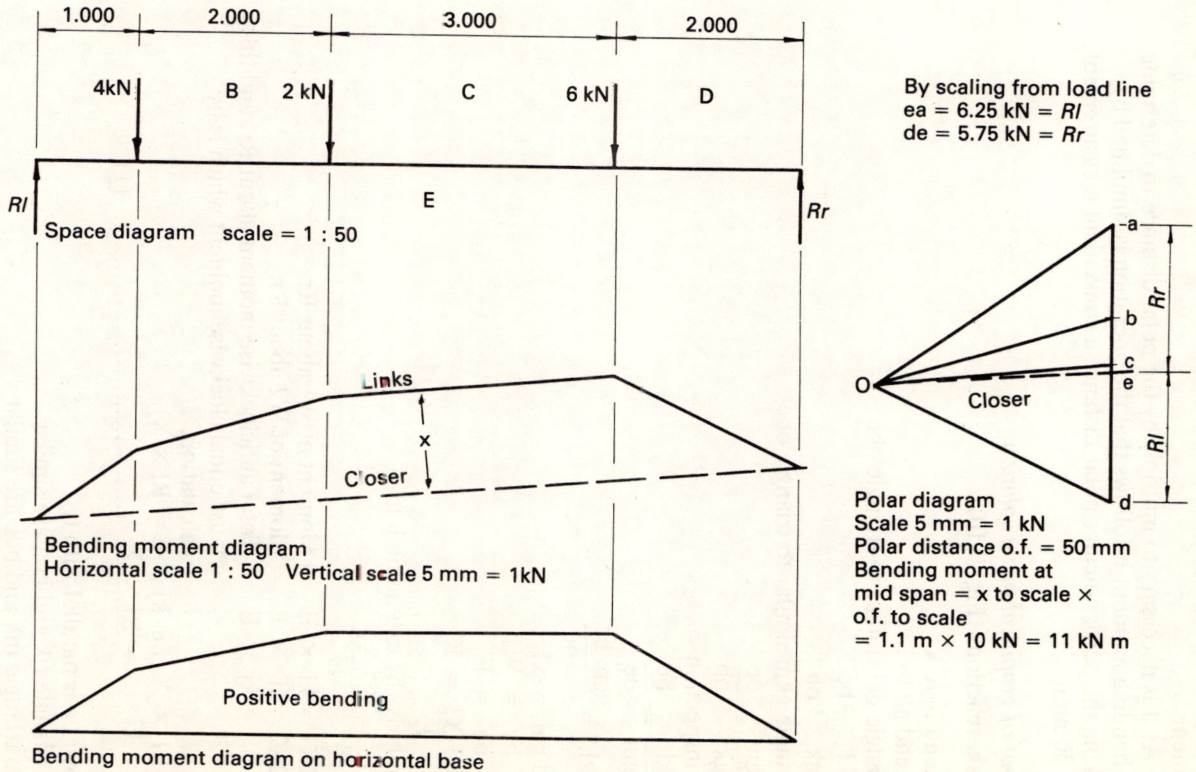

Fig. 9.9 Bending moment diagram by graphical method

To determine the bending moment value at any point on the beam, measure to scale the vertical distance enclosed by the link polygon and multiply this by the perpendicular polar distance to give the bending moment.

As it is necessary to multiply by the polar distance to determine the bending moment it follows that the operation is simplified by making the perpendicular polar distance a convenient measurement, i.e. 50 mm.

Proof of graphical beam solution

Beam reactions (Fig. 9.10)

Extend fg to k

Extend gh to l

Triangle gkl similar to triangle obc

$$\therefore \frac{kl}{gk} = \frac{bc}{ob}$$

Triangle gkm similar to triangle obe

$$\therefore \frac{gm}{gk} = \frac{oe}{ob}$$

By inspection

$$\frac{kl}{gm} = \frac{bc}{oe}$$

$$\therefore kl = \frac{gm \cdot bc}{oe}$$

and gm = x1

bc = W1

oe = E

$$\therefore \frac{w1\ X1}{E} = Kl$$

Similarly, by extending h to v

$$lj = \frac{W2\ X2}{E}$$

W1X1 = kl × E = Moment of W1 about Rr

W2X2 = lj × E = Moment of W2 about Rr

∴ kj × E = Total anticlockwise moment about Rr, and this must equal clockwise moment which is by calculation:

Rl × X (i.e. kj × E = Rl × X).

$$\therefore \quad kj = \frac{Rl \times X}{E} \qquad (1)$$

Now: oa is parallel with fj

∴ triangle fjk and boa are similar

also triangle fnk and boe are similar

$$\therefore \quad \frac{kj}{kf} = \frac{ba}{bo}$$

and

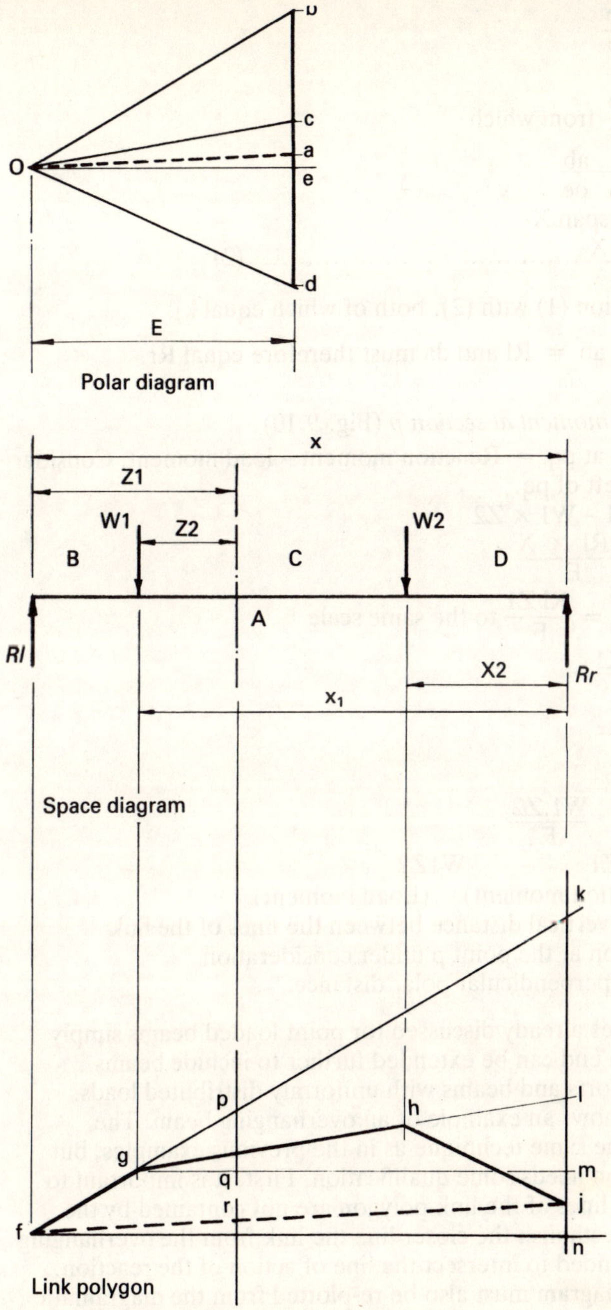

Fig. 9.10 Proof of graphical solution

$$\frac{kf}{fn} = \frac{ob}{oe}$$

and

$$\frac{kj}{fn} = \frac{ab}{oe} \text{ from which}$$

$$kj = \frac{fn\ .\ ab}{oe}$$

fn represents the span X

$$\therefore \quad kj = \frac{abX}{E} \quad \ldots\ldots\ldots\ldots\ldots\ldots\ldots\ldots \quad (2)$$

By relating equation (1) with (2), both of which equal kj,

$$\frac{ab.X}{E} = \frac{Rl.X}{E} \therefore ab = Rl \text{ and da must therefore equal Rr.}$$

Proof of bending moment at section p (Fig. 9.10)

Bending moment at pq = Reaction moment – load moment. Consider moments to the left of pq.

$$B.M. = Rl \times Z1 - W1 \times Z2$$

From (1) $kj = \dfrac{Rl \times X}{E}$

To some scale pr $= \dfrac{Rl\ Z1}{E}$ to the same scale

Since Kl $= \dfrac{W1X1}{E}$

$$pq = \frac{W1Z2}{E}$$

$$qr = pr - pq$$

$$qr = \frac{Rl.Z1}{E} - \frac{W1.Z2}{E}$$

$$qr \times E = Rl \times Z1 \quad - \quad W1Z2$$

(Reaction moment) (Load moment)

qr – is the vertical distance between the lines of the link polygon at the point p under consideration

E – is the perpendicular polar distance.

The techniques already discussed for point loaded beams simply supported at each end can be extended further to include beams overhanging supports and beams with uniformly distributed loads.

Figure 9.11 shows an example of an overhanging beam. The method follows the same technique as in the previous examples, but the major variation needs some qualification. First, it is important to note that the link lines of the link polygon are not contained by the reactions, and to establish the closer line the link from the overhanging load must be extended to intersect the line of action of the reaction. The shear force diagram must also be re-plotted from the diagramatic form to follow the convention of positive and negative shear values above and below the datum.

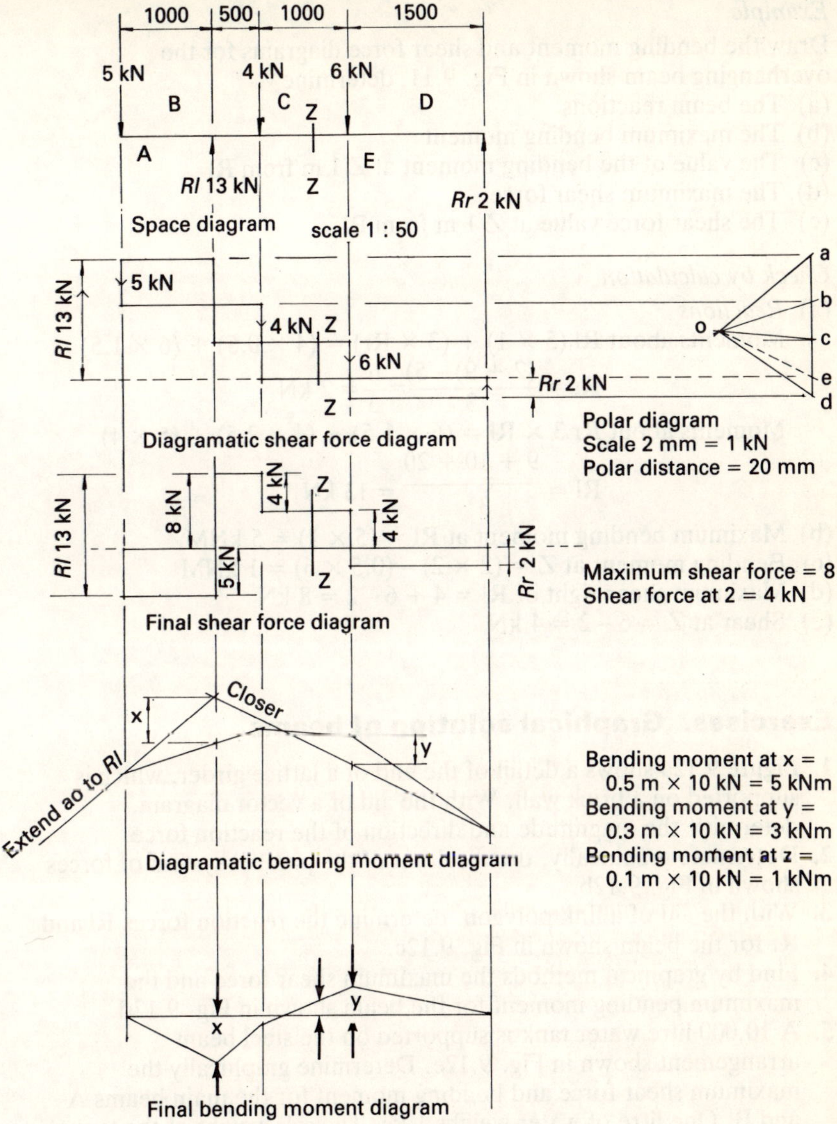

Fig. 9.11 Example of beam with overhanging end

Example

Draw the bending moment and shear force diagrams for the overhanging beam shown in Fig. 9.11, determine

(a) The beam reactions
(b) The maximum bending moment
(c) The value of the bending moment at Z 1 m from Rl
(d) The maximum shear force
(e) The shear force value at Z 1 m from Rl.

Check by calculation

(a) *Reactions*

Moments about Rl $(5 \times 1) + (3 \times Rr) = (4 \times 0.5) + (6 \times 1.5)$

$$Rr = \frac{(2 + 9) - 5)}{3} = \frac{6}{3} = 2 \text{ kN}$$

Moments about Rr $3 \times Rl = (6 \times 1.5) + (4 \times 2.5) + (5 \times 4)$

$$Rl = \frac{9 + 10 + 20}{3} = 13 \text{ kN}$$

(b) Maximum bending moment at Rl = $(5 \times 1) = 5$ kNM
(c) Bending moment at Z = $(2 \times 2) - (0.5 \times 6) = 1$ kNM
(d) Maximum shear right of Rl = $4 + 6 - 2 = 8$ kN
(e) Shear at Z = $6 - 2 = 4$ kN

Exercises. Graphical solution of beams

1. Figure 9.12a shows a detail of the end of a lattice girder, which is supported on a brick wall. With the aid of a vector diagram, determine the magnitude and direction of the reaction force.
2. Determine graphically, the resultant of the parallel system of forces shown in Fig. 9.12b.
3. With the aid of a link polygon, determine the reaction forces Rl and Rr for the beam shown in Fig. 9.12c.
4. Find by graphical methods the maximum shear force and the maximum bending moment for the beam shown in Fig. 9.12d.
5. A 10,000 litre water tank is supported on the steel beam arrangement shown in Fig. 9.12e. Determine graphically the maximum shear force and bending moment for the main beams A and B. One litre of water weighs 1 kg. The self weight of the tank and the beams may be disregarded.

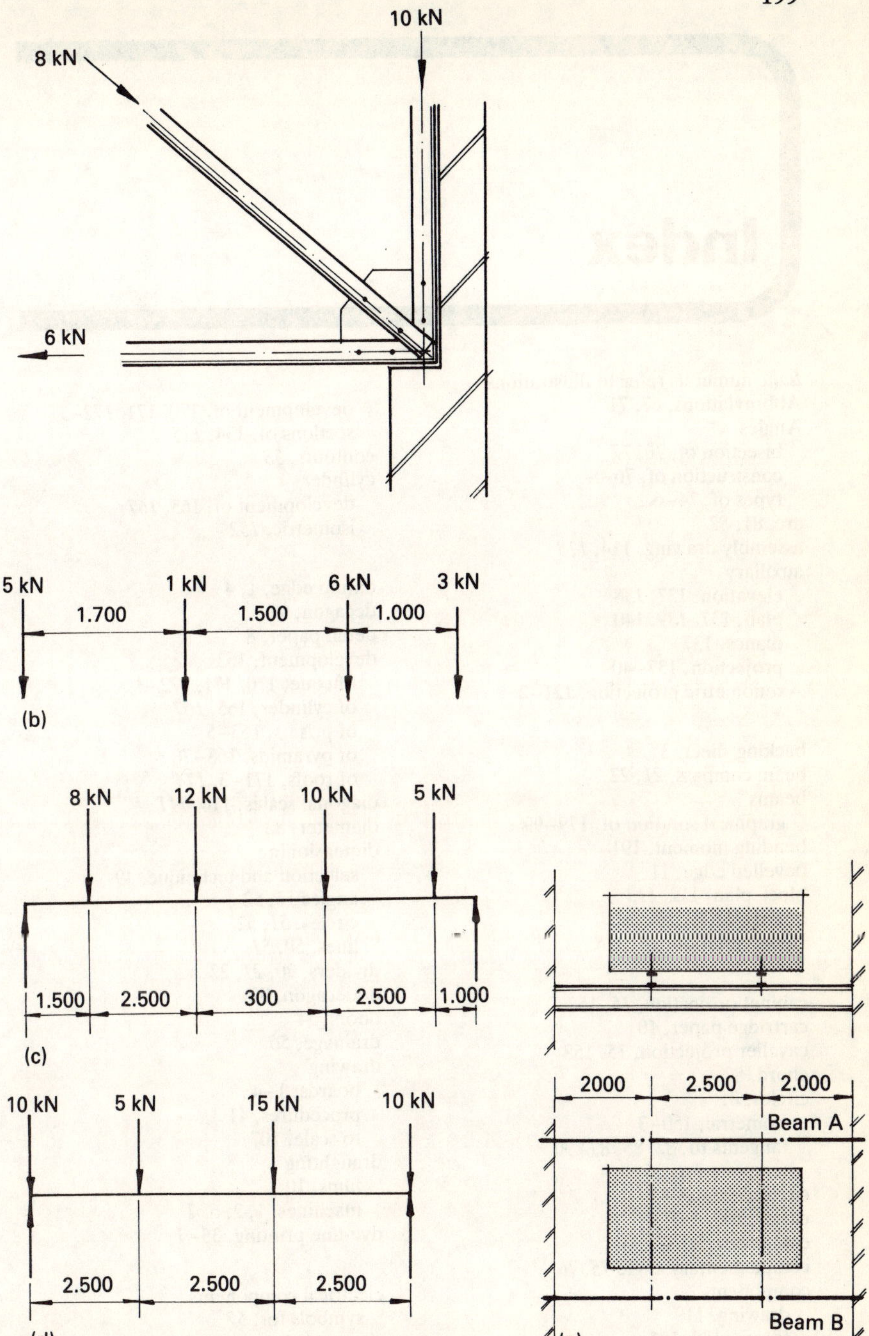
10 kN
8 kN
6 kN
5 kN
1 kN
6 kN
3 kN
1.700
1.500
1.000
(b)
8 kN
12 kN
10 kN
5 kN
1.500
2.500
300
2.500
1.000
(c)
10 kN
5 kN
15 kN
10 kN
2.500
2.500
2.500
(d)
2000
2.500
2.000
Beam A
Beam B
(e)

Fig. 9.12

Index

Italic numerals refer to illustrations
Abbreviations, 67, 71
Angles
bisection of, 76, *77*
construction of, 76–8
types of, 74–6
arc, 81, *82*
assembly drawing, 114, *118*
auxiliary
elevation, 137, *138*
plan, 137, *139*, 140
planes, 137
projection, 137–40
Axonometric projection, 121–2

backing sheet, 3
beam compass, *21*, 22
beams
graphical solution of, 179–99
bending moment, 191
bevelled edge, 11
block plan, 112, *113*
boards, 1–4
Bows Notation, 181, *183*

cabinet projection, *15*, 158
cartridge paper, 10
cavalier projection, *15*, 158
chord, 82
circles, 81
isometric, 150–3
tangents to, 82, 85, 87–90
template for, *26*
circumference, 81
closer, *190*, 191
compasses, 20
complementary angle *75*, 76
component
drawing, 119
forces, *180*, 181
symbols for *54–8*
cone
development of, 170, 171, *172–3*
sections of, 134, *135*
contours, *58*
cylinder
development of, 165, *167*
isometric, *152*

datum edge, 1, *4*
decagon, 92
detail paper, 8
development, 163
of cone, 170, 171, *172–3*
of cylinder, 165, *167*
of prisms, 163–5
of pyramids, 168–70
of roofs, 171–3, *174*
diagonal scales, 110, *111*
diameter, 82
dimensioning
selection and technique, 49
angles *51*, 52
circles, *51*, 52
lines, 50, *51*
dividers, 20, *21*, 22
dodecagon, 92
doors, *54*
drainage, *56*
drawing
boards, 1–4
procedures, 41
to scale, 107
draughting
films, 10
machines, 1, 2, 6, 7
dye-line printing, 35–7

electrical components
symbols for, *57*
electrostatic copying, 37–9
elevations
auxiliary, 137

building, 114
in first angle, 123–5
in third angle, 125–9
ellipse, 95
approximation to, 99, *100*
construction of, 97–9
normal to, 99–100
equilibrant, 179, *180*, 184, 185
eraser, 17–19
erasing shield, 18, *19*

film
adhesive, 29, *32–3*
polyester, 10
fire precautions, symbols for, *51*
flexible curve, 24
focal points, 95
folding drawings, 29, 34, *35*
forces
component of, *180*, 181
concurrent, 179, *180*
co-planer, 179, *180*
identification of, 181
non-concurrent, 179, *180*
parallel, 179, *180*
resolution of, *188*, 189
fountain drawing pen, 14–16
french curve, 24, *28*

guided pen lettering, 63, –4, *66–7*

heptagon, 90
hexagon, 90
hidden detail, *46*, 129
hip rafter, *142*, 143, 171–3, *174*
hyperbola, 101, *103*

inks, 16–17
instruments, 20–2
isometric projection, 148–55

layout, 47–9
lettering, 53, 60–9
linear scales, 108–10
lines
bisection of, 72–4
division of, 72, *73*, 74
linework, 43, –7
link polygon, 189–91

nib
types of, *59, 15*
holder, *59*
nonagon, 90
normal
to a circle, 82
to an ellipse, 99–100
north point, 53
notation, 67
numerals 62–3

oblique projection, 156–61
octagon, 90
Offset lithography, 39, *40*
orthographic projection, 122

paper
sizes of, 7, 8, *9*
types of, 8, 10
parabola, 101, *102*
parallel
lines, 46
motion, 1, 2, 5
system of forces, 179, *180*, 186
parallelogram draughting machine, 7
pens, 14–16, *59*
pencils, 12
clutch, *13*, 14
fine lead, *13*, 14
grades of, *13*
sharpeners for, *13*, 14
pentagon, 90
perpendicular bisector, 72, *73*
plan
auxiliary, 137
building,
sectional, 131, *132*
polygons
types of, 90–8
construction of, 92–5
link, 189–91
polyester film, 10
polar diagram, 186, *187*
prisms, development of, *164*, 165
projection
auxiliary, 137–40, *141*
axonometric, 121
cabinet, *157*, 158
cavalier, *157*, 158
first angle, 123–5
isometric, 148–55
multi-plane, 121–2
oblique, 156–61
planes of, 121–2
planometric, 158–60
third angle, 125, *127, 128*, 129

quadrant
of a circle, 82
first, 122–3
third, 122, 125, *127*

rabatment, 140
radius, 82
railway curve, 25
reactions, of beams, 189
reprographic processes, 34
resultant, 179, *180*, 184, *184*, *185*
roman lettering, 62, *60–1*
roof development, 171–3, *174*
ruling pen, 14–16
running dimensions, 50, *51*, *147*

sanitation, symbols for *56*
scale
diagonal, 110, *111*
drawing to, 107
linear, 108–10
of chords, 77–8
selection of, 110
sector, 82
sections, 129
building, *115*, *117–19*, 131, *133*, 134
conic, 134–5
sectional elevation, 130–1
sectional plan, 131, *132*
segment, 82
set squares, 11–12
sharpeners, *13*, 14
shear force, 191–2, 197–8
site plan,
spring bow, 20, *21*
stairways, *55*
stencil lettering, 63, *66*
straight edges, 4
survey, symbols for, *58*
symbols for drawings, 52–3, *54–8*

tangent, 82, 85
construction of, 87–90
tee squares, 1, 4
technical pen, 14–16
templates, 24, *26–7*
third angle projection, 125, *127*, *128*, 129
title blocks, *42*, 43
tracing paper, 8
tracing cloth, 10
track and trolley, 6
trammel, *98*, 99
transfers, pressure, 25
transition, 174, *176*
triangles, 78–81
true lengths, 13–37

undecagon, 92
Uno lettering guide, 63–4, *67*

vector
diagram, *183*, 184
line, 181, *182*
quantity, 181, *182*

Water supply, symbols for *56*
Windows, symbols for *54*